Applied Mineralogy
Technische Mineralogie

Edited by
Herausgegeben von

V. D. Fréchette, Alfred, N.Y.
H. Kirsch, Essen
L. B. Sand, Worcester, Mass.
F. Trojer, Leoben

3

Springer-Verlag
Wien New York 1972

D. A. Gerdeman, N. L. Hecht

Arc Plasma Technology in Materials Science

Springer-Verlag

Wien New York 1972

DENNIS A. GERDEMAN, B.M.E., M.S., Research Scientist,
University of Dayton Research Institute, Dayton, Ohio, U.S.A.

NORMAN L. HECHT, B.S., M.S., Ph. D., Research Ceramist,
University of Dayton Research Institute, Dayton, Ohio, U.S.A.

With 73 Figures

ISBN 3-211-81041-2 Springer-Verlag Wien-New York
ISBN 0-387-81041-2 Springer-Verlag New York-Wien

This book is dedicated to our parents,
Mr. and Mrs. Hyman Hecht and Mr. and Mrs. Felix Gerdeman

Preface

Although a considerable amount of information concerning the applications for arc plasmas in the materials sciences is available, it is contained in literally thousands of separate manuals, technical notes, textbooks, and government and industrial reports. Each source generally deals with only one specific application or, at best, a narrow range of utilization. This book was developed to provide a comprehensive and up-to-date compilation of information in the technology of arc plasma utilization. The book is divided into two general categories: flame spraying and materials evaluation.

In the flame spraying section a comprehensive review of the plasma spraying process is presented. The design and operation of plasma spraying equipment are described. Included are a description of the nature of a plasma, and the design and operation of plasma generators, powder feed systems and accessory control equipment. The general process procedures, and associated process variables are described. Particular emphasis is given to the particle heating process and the mechanisms for adherence and cohesion of coatings. Competitive flame spraying equipment is also detailed (combustion process, detonation and electric arc) and compared with the plasma spray process. A discussion and compilation of flame sprayed ceramic and metal materials, their properties and applications are also included.

In the chapters dealing with arc plasma testing, the various types of test facilities which utilize the arc plasma as an energy source are reviewed. The general areas of testing discussed include reentry simulation, thermal stress, thermal shock, ablation, dynamic oxidation, rocket exhaust simulation and rocket nozzle evaluations. The advantages and limitations of arc plasma testing are compared with competitive evaluation techniques. Diagnostic instrumentation, including heatflux meters, enthalpy probes, pitot tubes, ablation gauges, and devices for determining ionization level, etc., are described and discussed at length. A major portion of the text is concerned with the difficulties involved in defining the extremely high temperature, nonequilibrium test environment, measuring model response, and interpreting test results. The final section of this text contains a comprehensive bibliography of literature dealing with flame spraying and plasma arc testing.

The authors wish to acknowledge the efforts of the many individuals who have contributed in one way or another to the successful completion of this manuscript. In particular we wish to thank our editor, Dr. V. D. FRÉCHETTE for his many constructive comments. We also wish to thank Dr. W. G. LAWRENCE, of the State University of New York College of Ceramics, Alfred University, and Dr. L. I. BOEHMAN of the University of Dayton Research Institute for critiquing these works. The assistance provided by Mrs. JUDITH HECHT in reading the manuscript and compiling the bibliography is also gratefully acknowledged. The efforts of Mr. FRED KRAMER in preparing many of the illustrations, Mrs. CARMA MERKERT in checking the references, Mrs. MARY ROSENBERG in checking the index, and Mrs. GAIL ST. FELIX MOORE in typing and retyping the manuscript are likewise acknowledged.

Finally the encouragement, understanding, and phenomenal patience of our families, and especially our wives, JUDITH HECHT and JANE GERDEMAN, are appreciated.

Dayton, Ohio, Summer 1972   DENNIS GERDEMAN
                                                       NORMAN HECHT

Table of Contents

1. Introduction

The arc plasma has achieved significant importance in a diversity of research and industrial applications within the past twenty years. Of major interest is the use of the arc plasma for flame spraying and as a heat source for material testing. The range of plasma arc applications is illustrated in Fig. 1.

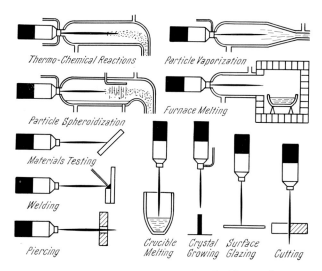

Fig. 1. Industrial Applications for the Plasma Arc
(after Thermal Dynamics Corp.)

The flame spray process is one of the major methods employed in the fabrication of coatings and thin-walled components. In this process materials are introduced into a hot gas stream (effluent) and propelled onto the surface of a substrate or mandrel. The heated particles are generally in a molten or plastic state and are rapidly cooled upon impact on the cooler surface. The impacting particles flatten, interlock and overlap one another, securely bonding together and forming a coherent layer of material. When the substrate is properly prepared, an adherent bond is formed between the coating layer and the substrate.

There are three principal flame spray processes:

(1) Plasma Spraying,
(2) Combustion,
(3) Arc Spraying.

In addition there are a number of variations in each process depending on the form in which the material is introduced into the effluent, the gas pressure, the combustion rate, etc. The primary emphasis in this text is on the plasma spray process.

The flame spray process dates back to the early 1900's. The invention of the metallizing process around 1910 is credited to MAX ULRICK SCHOOP of Switzerland. In the Schoop process a combustion-heated gun is employed for spraying both metal and ceramic materials. The major application was spraying zinc coatings for corrosion resistance. By the 1930's the applications had been extended to the spraying of hard metals to build up machine parts. With the advent of the space age in the 1950's greater emphasis was placed on the development of new materials and improved processing for coating applications. In the latter half of the 1950's the plasma torch was introduced to the flame process to meet aerospace requirements.

The development of the plasma torch has provided the coating technologist with a research tool for investigating and developing new and improved coatings. It provides a controlled environment with temperatures in excess of the melting points of all known materials. For the past few years an extensive effort has been expended in the investigation and development of plasma-sprayed coatings and considerable progress has been made in this field.

The advent of the space age likewise called for techniques capable of evaluating and aiding the improvement of newly developed high-temperature materials needed for rocket engines, jet engines and hypersonic flight vehicles. Methods were required which would be capable of duplicating, in a laboratory, the conditions which a space vehicle would encounter during reentry into the earth's atmosphere. In addition, test facilities to study the aerodynamics of hypervelocity flight were required. The most important item necessary for the establishment of the needed test facilities was a source of high-energy gas. This requirement is most appropriately satisfied by the arc plasma, which has achieved great success for this and a wide variety of related uses. The plasma jet is a valuable tool because of its versatility, wide range of operating conditions which can be closely controlled (both thermodynamically and chemically), and the extremely high effluent temperatures and enthalpies which can be attained.

Plasma facilities have been established to conduct a variety of materials evaluations. Tests have been designed to evaluate ablative-type materials concepts, thermal-shock and thermal-stress characteristics and to conduct dynamic oxidation studies. More elaborate facilities are capable of simulating vehicle reentry or the exhaust of a rocket engine. Many of the large number of facilities which were established in government installations, private industry and educational institutions were built specifically to meet the requirements of the individual organization and to solve problems of immediate concern. The instrumentation used in conjunction with these test facilities was likewise often designed to serve individual or highly specialized needs.

All of the early test facilities contained 40- to 50-kilowatt units which operated on nitrogen and produced enthalpies between 12,000 and 16,000 Btu per lb. After mixing with oxygen to simulate air, enthalpies dropped to the 8,000 to 12,000 Btu per lb range. These facilities were capable of producing the high heat-transfer rates encountered on ICBM nose cones. The small units were soon replaced by 100- and 200-kilowatt plasma jets which could accommodate larger test sections than their predecessors. The required continuous operating time was also increased as interest grew in reentering orbital vehicles. With a few exceptions the plasma test facilities of the 1950's were limited to less than five-minute runs and enthalpies of about 15,000 Btu/lb. The past decade however, has produced sizable improvements in the enthalpies attainable in small units and also has resulted in the development of much larger units which are capable of operating at considerably higher arc-chamber pressures. Units in the megawatt range are quite common and at least one installation has the capability of operating at 50 megawatts.

The high enthalpy, high-velocity effluent which is produced must be well defined before it is really useful. A considerable amount of time and effort has been expended on the development of suitable instrumentation to be used in these facilities. Ideally, when the test conditions can be accurately defined the performance of the test model permits prediction of the performance to be expected in actual flight.

The plasma torch is also an important tool in a number of material processing applications. The high-temperature plasma flame can be used for cutting metals, for beneficiating raw materials, and for specialized processing of powders, and it can serve as a heat source in furnaces. Powders injected into the hot effluent can be easily spheroidized. By rapid quenching, materials can be retained in the glassy state and high-temperature phases can be preserved. The unusual phases encountered in materials obtained by this rapid quenching of plasma sprayed powders are of considerable mineralogical interest.

2. The Plasma Arc

2.1. The Plasma State

Plasma is a gaseous cloud composed of free electrons, positive ions, neutral atoms and molecules. It has been referred to as "the 4th state of matter" because of its unique properties. Its behavior involves complex interactions between electromagnetic and mechanical forces.

A plasma is generated whenever sufficient energy is imparted to a gas to cause at least partial ionization. However, the plasma domain really begins above $20,000°$ K, where an appreciable percentage of the atoms are ionized. The plasma is initiated when electrons are accelerated between two electrodes in a gaseous environment. As the electrons speed toward the anode they collide with and excite the atoms or molecules in the gas. This excitation can cause complete ionization, orbital displacement of an electron or merely increased kinetic energy. The additional electrons freed by ionization are also accelerated toward the anode and in turn cause more collisons and further ionization. The result is a highly conductive gas through which current in the form of an electrical spark can pass. Sparking between electrodes results in a greatly increased current flow and electron emission. Collisions between electrons and the larger particles become more frequent; these collisions transfer the kinetic energy of the electron to the particles and raise the temperature of the gas. It takes only a little energy to make the electrons very "hot". The acceleration of an electron through a voltage drop of one volt corresponds to an approximate temperature of $20,000°$ K. The other particles in a plasma are heated by colliding with electrons. It is more difficult to raise the kinetic energy of particles to a level corresponding to that of the electrons because of their greater mass, and therefore a greater frequency of collisions is necessary. One means of raising the kinetic energy of particles is by increasing the gas pressure of the system. When the gas particles attain the same kinetic energy level as the electrons the plasma has attained equilibrium for a given electrical energy input. Thermal and magnetohydrodynamic effects are also employed to increase the pressure and thereby the density of the arc. As the conductivity of the plasma increases, the number of electrons bombarding the anode also increases, thereby raising the temperature of the anode and freeing from the surface ions, which are accelerated toward the cathode.

The temperatures within the plasma effluent can vary from 7,000 to $10,000°$ K. The electrons in the plasma will generally be in the $100,000°$ K range while the other particles (ions, atoms and molecules) will be about $10,000°$ K. The plasma has a high electron current density and behaves as a dielectric in an electric field, and as a diamagnetic in the presence of a magnetic field.

To maintain the plasma state, energy must be added to make up for conduction and radiation losses. However, an increase in electrode current can cause the voltage requirements to go through an erratic history of change. Voltage decreases as the current is increased, until a stable condition is obtained at a high current-low voltage condition. A profile of the voltage traverse across the arc is shown in Fig. 2. As can be seen, there is a low voltage drop at the negative electrode [1, 2].

Although a high-pressure plasma can be easily obtained within a closed container it has limited value. It is more desirable to have a high-pressure plasma of minimum confinement. High-pressure plasma generators of minimum confinement can be

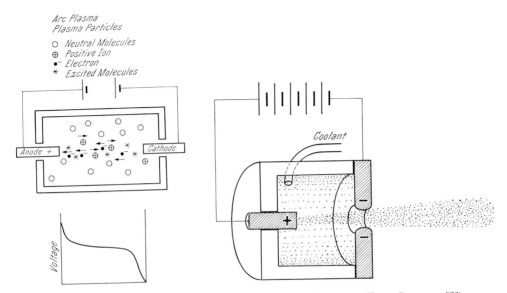

Fig. 2. Schematic Representation of the Conventional Arc and Its Characteristic Voltage Gradient (From HELLUND [1])

Fig. 3. Plasma Generator (From GIANNINI [2])

designed if the naturally occurring thermal and magnetohydrodynamic forces are employed. This type of design generally consists of a small cylindrical chamber (glass or metal) in which the arc is struck. The front end of the chamber has an orifice for the emerging effluent. This front wall of the chamber also acts as one of the electrodes. The second electrode is generally located at the back of the chamber. Water or gas is injected tangentially into the chamber (see Fig. 3). The arc is contained within a fluid vessel which acts as a high-pressure cushion between the arc and the walls of the chamber. This fluid vessel cools both the chamber and the gas in the outer regions of the plasma, thereby lowering the ionization and conductivity of the gas in the outer region. This causes the discharge current to concentrate in the central zone of the effluent, thus increasing the density, temperature and conductivity of this zone of the effluent.

This phenomenon is commonly referred to as the "thermal pinch effect". As the current is increased due to the thermal-pinch effect, the self-induced magnetic field around the plasma also increases and a "magnetic-pinch effect" occurs. This is due to the fact that when current flows through parallel conductors in the same direction the conductors will be attracted to each other by their self-induced magnetic fields. The charged particles in the plasma jet behave in a similar manner (note that the kinetic gas pressure acts in opposition to the magnetic forces, but these forces are much weaker and have little effect).

As the effluent emerges from the orifice it gives up energy to its surroundings. Part of this energy is dissipated as heat by conduction and radiation. There is also kinetic energy contained in the directed movement of the mass of the plasma as a whole, which is also dissipated as heat. In addition, energy is dissipated when the ions recapture their electrons or the dissociation products reunite. A significant component of the radiation from the plasma is generated by electron-electron collision. This high-temperature plasma jet provides an ideal means for melting and propelling particles of materials onto the surface of a substrate, as well as for simulating various high-temperature environments [2, 3, 4, 5, 6].

2.2. Plasma Generators

A plasma can be generated either with or without the use of electrodes as shown in Fig. 4. However, the arc-plasma generators with electrodes are of primary interest for flame spraying and material testing. There are two major types of arc plasma generators: the nontransferred arc and the transferred arc. In the nontransferred arc both electrodes are contained in the chamber, as is the arc through which the effluent passes before it emerges from the orifice in the anode. In the transferred arc the arc is struck through a hollow nozzle to an external conducting member. The external member is usually the workpiece to be heated. This type of device is used for welding, cutting, or other processing requiring high workpiece temperature.

The *nontransferred-arc plasma generators* can be further subdivided into five categories: vortex stabilized, gas-sheath stabilized, wall stabilized, magnetically stabilized and water stabilized. The vortex or the gas-sheath stabilized designs are generally used for flame spraying, while all five are used for material testing. More recently the use of the induction plasma has also been reported [7, 8].

In the *vortex-stabilized* unit, gas is fed tangentially into the chamber between electrodes to produce an intense vortex off the tip of the rear electrode, with a corresponding low-pressure area at the center of the vortex. The arc is stabilized, or contained, in the vortex and travels down the front electrode. The tangential flow of the gas provides a cool boundary layer surrounding the plasma jet, reducing heat transfer to the front electrode. The electrodes are made of tungsten, carbon or copper and are generally water-cooled.

In the *gas-sheath stabilized* plasma jet, the arc gas is introduced from behind the rear electrode. The arc path is between a solid tungsten cathode to a hollow water-cooled copper anode. The arc remains in the nozzle and is prevented from contacting the wall by a gas sheath much thicker than the arc diameter. However, the arc is allowed to strike through the gas sheath after passing a considerable distance down the nozzle. Arc positioning is accomplished by the gas flow pattern and control of turbulence.

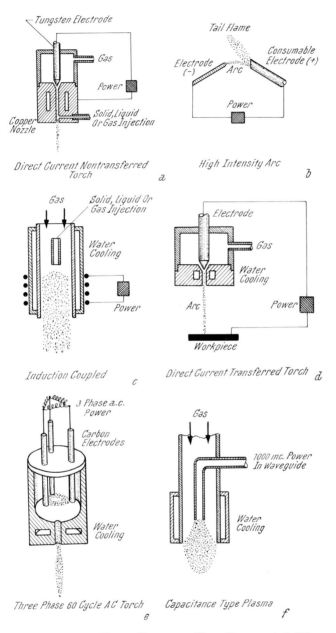

Fig. 4. Types of Plasma Generators (From Dennis *et al.* [3])
(a) Direct-Current Nontransferred Torch
(b) High-Intensity Arc
(c) Induction-Coupled Arc
(d) Direct-Current Transferred Torch
(e) Three-Phase 60-Cycle AC Torch
(f) Capacitance-Type Plasma

The *wall-stabilized* design consists of one solid and one hollow electrode with a geometry such that the plasma gas, instead of constricting the arc, becomes an integral part of the arc stream, filling the nozzle from wall to wall.

In the *magnetically stabilized* jet, the arc strikes radially from an inner to an outer electrode. Gas is blown axially through the annulus. The arc is rapidly rotated with a magnetic field so that its position at various times resembles a spoke in a wheel. This design permits operation at high pressures without electrode erosion.

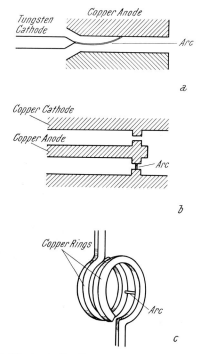

Fig. 5. D. C. Electrode Configurations (From DENNIS *et al.* [3])

In the *water-stabilized* generator both electrodes are consumable and the solid electrode is fed either manually or automatically to keep arc length constant. Water is swirled in and taken out the back end in some designs, while in others the water and gas both leave the nozzle. Consumption of electrodes is high and the plasma produced is generally highly contaminated.

The three basic electrode materials commonly employed in plasma generators are tungsten, copper and graphite. For most applications tungsten is used as a cathode with copper as the anode. Tungsten or tungsten with small additions of thoria or zirconia is used because of its good thermionic emission. Copper is used because of its high thermal conductivity. When copper is used, some mechanism must be supplied to rotate the arc foot point on the surface of the electrode to prevent local overheating and melting. For low currents, wall stabilization or the development of a cold gas sheath along the electrode surface is sufficient to protect the anode. A better method of assuring arc rotation is the use of a vortex motion in the arc gas. For high current the arc is rotated by means of a magnetic field.

The other major electrode material is graphite. The high vaporization temperature of this material precludes the necessity for cooling or for electrode rotation. The major disadvantage in the use of graphite lies in the resulting carbon contamination of the effluent. Some carbide electrodes have also been tried, without a great deal of success.

The configuration of the electrodes is very important in the operation of the plasma generator. An improperly designed set of electrodes will result in erratic arcing and unstable operation. The most common configuration employs one stick electrode and one annular or cylindrical electrode. The gas enters around the stick and after passing through the arc exhausts through the hollow electrode. The cylindrical electrode constricts the arc by forcing it to a smaller diameter. When tungsten is used, the stick is the cathode, and the cylinder, generally copper, is the anode. When graphite is used, the stick is generally the anode. The polarity of the stick is determined by which electrode must dissipate the lesser amount of heat. Two other configurations employed are the concentric rings (solenoid arc) and the parallel rings. All three D.C. electrode configurations are shown in Fig. 5 [2, 3, 4, 5, 6].

3. Plasma Spraying

3.1. Plasma Spraying Equipment

As previously stated, both the gas-sheath stabilized plasma jet and the vortex-stabilized plasma jet are used in the plasma spray process. The gases that have been most commonly used in the generation of a plasma arc are hydrogen, helium, nitrogen, argon and air. Gas selection is based primarily on gas energy, reactivity and cost. The relationship between plasma temperature and gas energy content is shown in Fig. 6 [9]. From Fig. 6, it can be seen that the energy content of nitro-

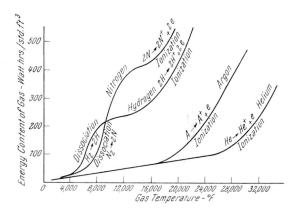

Fig. 6. Plasma Temperature as a Function of Gas Energy Content at Atmospheric Pressure
(From INGHAM and SHEPARD [9])

gen and hydrogen is considerably higher than that for argon or helium, due to the dissociation reactions in the nitrogen and hydrogen prior to ionization. The low cost and high internal energy of nitrogen makes it the most commonly used arc gas. Generally about 10% hydrogen is mixed with the nitrogen to increase the heat content and improve the heat transfer characteristics of the arc gas. Hydrogen also acts as a reducing agent in the plasma. Reactions of the nitrogen gas with the material being sprayed (nitride formation) somewhat limits the use of nitrogen gas. If a completely inert atmosphere is required, argon is usually selected as the arc gas. The use of air is very limited because of the excessive electrode oxidation. Excessive electrode wear, encountered when pure nitrogen is the arc gas, has been reduced by winding a coil of copper tubing concentrically about the nozzle axis in series with the arc power. This introduces a spinning component to the electric field, which maintains a fully annular arc, resulting in uniform arc impingement around the front nozzle. The copper coil does not alter the spray characteristics but greatly increases nozzle life [10].

The arc gas flow rate and the electric power of the plasma torch must be properly balanced. Gas flow must be carefully metered as the current is built up in the arc stream, so as not to blow out the arc nor fail to cause the necessary thermal pinch effect to force the arc down through the nozzle. Improper sequencing

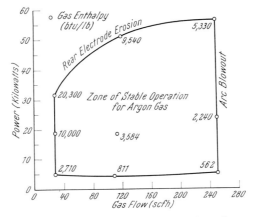

Fig. 7. Arc-Gas "Envelope" (After Plasmadyne Corp. [11])

can cause catastrophic failure of the nozzle and the electrode. The zone of stable arc operation is defined by an arc gas flow rate-arc power "envelope". A typical envelope for a plasma torch using argon as the arc gas is shown in Fig. 7. Gas enthalpies for the perimeter of the envelope are also shown.

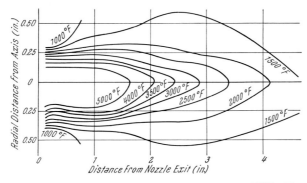

Fig. 8. Temperature Distribution in Arc-Plasma Discharge (18 KW) (From SMITH [6])

For general plasma spraying the arc-gas flow rate will vary between 100 and 200 SCFH and the arc power will vary from 5 to 40 kilowatts. The power supply generally consists of three-phase transformers in conjunction with three-phase full-wave bridge rectifiers (selenium rectifier welders). The typical profiles for temperature distribution and velocity pressure distribution obtained in an arc-plasma discharge are shown in Fig. 8 and 9 respectively [6]. Thermal efficiencies (thermal energy output vs. electrical energy input) normally range between 50 and 70%. Plasma arc torches which will operate at 50 to 100 kilowatt and at 3 to 4 atmospheres are currently under development for spray applications. These high gas velocity guns should prove to be more efficient and provide improved coatings.

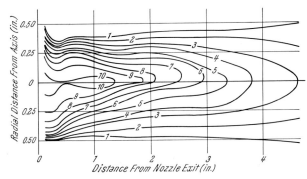

Fig. 9. Velocity Distribution in Arc-Plasma Discharge (18 KW) (From SMITH [6])

A variety of techniques are used for introducing material into the plasma effluent. The technique most widely used is powder injection. However, techniques for rod and wire injection have also been developed. A typical design for a powder plasma spray gun is shown in Fig. 10 [9]. Generally the powder is injected through a port in the front electrode. If extended particle dwell time is necessary (as with refractory metals or ceramics) the powder injection port may be inclined so that the powder is injected toward the rear electrode, or the powder can be injected through the rear electrode [12]. Attachments for injecting the powder beyond the front electrode are used for powders which require very short dwell time (plastics, low-melting metals). Front electrodes having a number of injection ports for spraying a mixture of powders are also available. It should be noted that powder injection into the vortex-stabilized jet is more critical than in the sheath-stabilized jet because of the swirling nature of the gas pattern in the vortex-stabilized jet. It is also important that the powder be injected in such a manner that it travels in the hot center zone of the effluent. This is an advantage of injecting the powder through the rear electrode; however, this generally results in severe erosion of this electrode.

The spray powder is stored in a hopper and transported to the plasma torch suspended in a carrier gas, usually the same gas as is used to generate the plasma arc. The powder is fed into the gas in controlled amounts by either an aspirator or an auger feed system. The hopper and feed system are usually mounted on a vibrational unit to prevent powder clogging. Small amounts of the carrier gas are generally passed through the hopper to fluidize the power and improve its flow properties.

Minimum carrier-gas flow rate is desirable so as to minimize cooling effects on the arc gas. Particle size distribution and powder feed rate are the two most critical parameters in the powder spray process. These two variables must be carefully controlled in order to achieve maximum particle melting.

A schematic of the powder spray process is shown in Fig. 11. The control console shown in Fig. 11 provides the controls for metering the gas, water and electric current flow to the plasma gun. The powder feed system generally has its own control system.

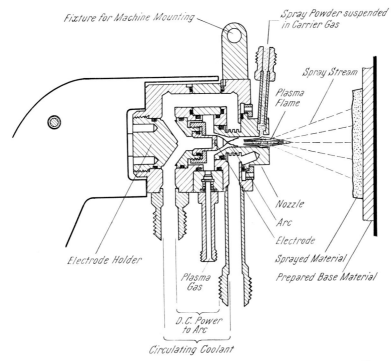

Fig. 10. Cross-Section of a Plasma Flame Spray Gun (METCO Type 2 MB [9])

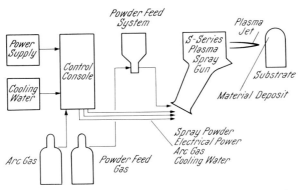

Fig. 11. Plasma Spray System (Plasmadyne S. Series [11])

Material injection by rod or wire eliminates the need for a carrier gas, since the feed rate is primarily controlled by the melting rate of the material at the tip of the rod or wire. A more uniformly sized particle is obtained by this process. Generally the rod or wire is fed into the plasma flame at a fixed rate and as the tip of the rod or wire melts it is carried into the effluent and propelled towards the substrate. The major advantage of this technique is that the particles of material enter the effluent in a molten state rather than having to be melted as they are propelled toward the substrate. The major limitation of the rod and wire processes is that only those materials which can be fabricated into a rod or wire form can be sprayed. Although designs for rod- and wire-fed plasma spray systems have been reported in the literature [13, 14, 15] they have not been marketed commercially.

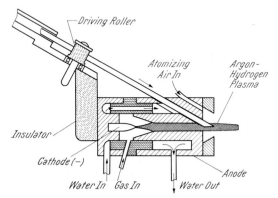

Fig. 12. Experimental Rod-Type Plasma Spray System
(From JOHNSON and WHEILDON [13])

Fig. 12 shows an experimental design for a plasma rod-spray system. Preliminary studies for this system, using a zirconia rod, showed that increased spray rates and greater deposition efficiency (percentage of consumed material which is deposited) could be obtained. Spray rates in excess of 6 lbs. per hr. at deposit efficiencies of 75% were obtained, as opposed to 2 to 4 lbs. per hr. at deposit effi-

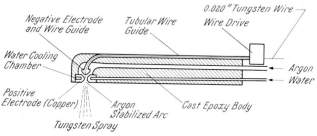

Fig. 13. Schematic of Wire-Type Arc Spray Gun
(From LEVINSTEIN et al. [15])

ciencies of 65% for conventional powder spray systems. It was also found that a shorter heat zone was required since only the tip of the rod had to be heated [14]. Ceramic rods for spraying can be fabricated by standard extrusion and sintering techniques or by bonding the fine particles with organic resins which burn out during spraying.

A schematic diagram of a wire-fed arc spraying gun is shown in Fig. 13. The 0.40-inch diameter wire introduced through the rear housing acts as a consumable rear electrode. A high-frequency arc starter initiates an arc which forms between the wire and the water-cooled front electrode. The arc is stabilized by the flow of highly pressurized argon gas. The gun operates at 40 kilowatts with a deposition rate of approximately 5 lbs. per hr. [15]. The nature of the process limits the materials which can be sprayed primarily to metals having good thermionic emission capabilities.

A liquid-stabilized plasma jet for spraying oxides and carbides has also been described in the literature [16]. Water was used for the oxides and a mixture of water and alcohol was used for carbide deposition. Deposit efficiencies of 74% were reported. The effects of electrode contamination in the sprayed coatings was not discussed.

3.2. The Plasma Spray Process (Powder Spraying)

This discussion of the plasma spray process is limited primarily to powder spraying. This has been done because most of the commercial and experimental work has been expended on the powder process. The ready availability of most materials in powder form and the fact that commercial equipment is designed for powder spraying explains why the vast majority of plasma spraying is being accomplished by the powder process.

A wide variety of metallic, ceramic and organic powders have been sprayed by the plasma process. The major criterion for spraying any material is that it have a distinct range over which it remains in the liquid phase. Incongruently melting materials can be sprayed provided that the individual constituents have a liquid phase. Materials which sublime cannot be sprayed, and those materials which have a very high viscosity as liquids are difficult to spray.

As seen from the schematic for the powder spray process (Fig. 11) a powder feed system is joined to a plasma spray system to deposit material on a substrate. The complete process includes: various prespray treatments for the powder and the substrate, the spraying step which incorporates powder, plasma and substrate and post-spray treatments of the sprayed composite. The complex nature of the spray process results in a large number of processing parameters which must be controlled in order to obtain optimum coatings. The process parameters may be categorized as in Table 1.

Although the spray process has a large number of associated parameters only a limited number of these can be closely controlled. To a large extent the parameters are fixed, based on the selection of spray material, substrate material and spray equipment. Generally only the surface condition of the substrate, the physical nature of the powder, the spray environment, and equipment variables can be regulated. Arc configuration and feed system are predetermined by equipment selection but can often be modified by accessories. It is important to note that electrode geometry changes with use and that the gas flow patterns will be altered by electrode wear. The properties of the plasma effluent can be somewhat regulated by gas flow rate and power level adjustments but the specific gas composition greatly effects the effluent behavior.

Although the properties of the spray material and substrate material are fixed, they must be considered prior to material selection. Incompatibility between

Table 1. *Summary of Plasma Spray Process Parameters*

I. Plasma

    A. Arc Configuration (plasma gun design)

        1. Electrode geometry
        2. Arc gas flow pattern (vortex stabilized vs. sheath or wall stabilized)

    B. Properties of the Plasma Effluent

        1. Arc gas velocity and pressure
        2. Arc gas enthalpy (temperature)
        3. Arc gas composition
        4. Chemical properties of the arc gas (reactivity)
        5. Thermal properties of the arc gas (conductivity)

    C. Plasma Spray Equipment Variables

        1. Plasma gun
            (a) Gun traverse rate
            (b) Spray angle
            (c) Spray distance (gun to substrate)
        2. Console control
            (a) Arc gas flow rate
            (b) Power level (KW input)
            (c) Cooling-water flow rate
        3. Spray environment (atmosphere-control chamber, gas shrouds, air, etc.)
            (a) Atmosphere
            (b) Pressure
            (c) Temperature
            (d) Substrate holding fixture

II. Powder

    A. Chemical composition

        1. Crystal phases
        2. Crystal structure (lattice dimensions)
        3. Impurities

    B. Physical nature

        1. Shape
        2. Density
        3. Size (powder particle size distribution)
        4. Microstructure (porosity)
        5. Surface nature
            (a) Surface films (oxide, moisture)
            (b) Surface finish

    C. Properties

        1. Thermal
            (a) Melting point (melting behavior)
            (b) Thermal conductivity
            (c) Thermal expansion
            (d) Specific heat
            (e) Heat of fusion

Table 1 *(continued)*

(f) Heat of vaporization
(g) Emissivity

2. Chemical
   (a) Reactivity with arc or carrier gas
   (b) Reactivity with substrate

3. Liquid State
   (a) Viscosity profile
   (b) Surface tension

D. Powder-feed system

1. Feed equipment (aspirator or screw feed)

2. Feed rate
   (a) Carrier-gas flow rate
   (b) Powder injection rate (volume or weight)

3. Injection mode into plasma effluent
   (a) Point of injection
   (b) Angle of injection

III. Substrate

A. Chemical Composition
   1. Crystal structure (lattice constants)
   2. Crystal phases

B. Properties

1. Thermal
   (a) Melting point
   (b) Thermal expansion
   (c) Thermal conductivity
   (d) Specific heat
   (e) Heat of fusion
   (f) Heat of vaporization

2. Physical
   (a) Hardness
   (b) Density

3. Chemical
   (a) Reactivity with powder (phase formation, solubility, etc.)
   (b) Reactivity with environment (arc gas, carrier gas, etc.)

C. Surface condition

1. Surface finish
2. Surface microstructure (porosity, etc.)
3. Surface contamination
   (a) Surface films (water, oxides, etc.)
   (b) Contamination (dirt, etc.)
4. Surface stresses
5. Surface temperature
6. Surface defects

substrate and spray material will result in poor coatings. Sometimes an intermediate material may be deposited on a substrate to offset property differences
between the substrate and the top coating layer. Two major uses for intermediate
coating layers are to improve adherence and to balance mismatched thermal
expansion coefficients.

Under ideal conditions complete control of all the variable spray parameters
should yield optimized coatings at maximum deposition efficiency. The specific
effects of the equipment variables have been studied using deposition efficiency
as a measurement gauge, and the results are shown in Fig. 14 [17]. From this study

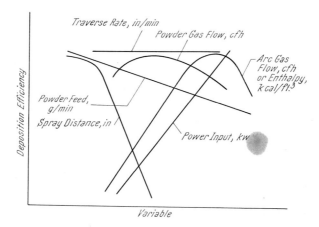

Fig. 14. Effect of Selected Variables on Deposition Efficiency
(From MASH *et al.* [17])

it appears that arc-gas flow rate, arc power level and spray distance are the most
critical variables. It is therefore necessary that these variables be properly regulated in order to provide the optimum combination of effluent enthalpy and
particle dwell time needed for obtaining particles which are completely molten
upon substrate impact.

For any specific arc gas the effluent pressure, velocity and enthalpy are
controlled by the combined effects of arc-gas flow rate and input power level.
Within the envelope of stable operation these two variables are jointly regulated
to accomodate the different thermal properties of the various spray materials.
Deposition efficiency increases as input power increases, as long as the power
level is maintained in the zone of stable operation, because increased input power
levels result in higher gas enthalpy and improved thermal conductivity.

Effluent velocity is one of the major factors controlling the velocity of the
powder particles. High particle velocity generally results in a higher density
coating and improved deposition efficiency. However, particle velocity should not
be so high as to prevent complete particle melting prior to impact. This effect is
illustrated in Fig. 14 where arc-gas flow rate reaches a maximum deposition
efficiency and then decreases with further increases in gas flow rate.

The effects of proper *particle dwell time* are also reflected in the curve for spray
distance versus deposition efficiency in Fig. 14. If the plasma torch is placed too

close to the substrate, the powder has too short a dwell time and will not be completely molten on substrate impact. In addition, too short a spray distance
generally results in overheating the substrate. Too large a distance is also undesirable, since extended particle dwell time results in resolidification of the particle
prior to impact. In general spray distances will vary between two and six inches.

In most spraying operations the effluent and spray powder impinge on the
surface of the substrate at a *spray angle* of 90°. Spray angle variation between
90° and 45° have no noticeable effect on the deposited coating. However, as the
spray angle decreases below 45°, shadow effects are observed in the coating
structure. Shadow effects are due to the buildup of particles on the substrate
surface which obstruct the flight path of subsequent particles and prevent the

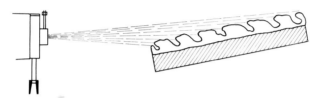

Fig. 15. "Shadow Effect", Caused by Low-Angle Spraying
(From INGHAM and SHEPARD [9])

filling of surface depressions. The effect is illustrated in Fig. 15. In addition to
causing a very porous structure, shadow effects also result in a weaker and less
homogeneous coating [9].

The *traverse speed* of the plasma gun is not a critical parameter but it should be
adjusted to permit a uniform rate of material deposition. In addition it is desirable to adjust the traverse rate so that the coating will consist of a number of
sprayed layers (maximum of 0.001 inch per pass). This reduces stress effects due
to coating shrinkage [9]. The traverse rate should also be fast enough to prevent
overheating of the substrate.

The physical nature of the powder prior to spraying can be controlled, to a
moderate degree, by various prespray treatments. It is important to closely control
the particle size and the size distribution of the powder. The particle size distribution range should be very narrow so that all the particles will require approximately the same thermal treatment for melting. In addition, a powder having a wide
particle size distribution range will be subject to a great deal of segregation in the
effluent prior to impact and this will result in a nonuniform coating layer. Too
coarse or too fine a particle size is also undesirable. The very fine particles have a
tendency to float on the periphery of the effluent while the very coarse particles
require excessive thermal energy for melting. Unmelted particles generally do not
adhere to the substrate on impact but rather rebound from the substrate. Those unmelted particles that are trapped on the substrate by the flow of surrounding molten particles become defective zones in the deposited coating.

The size, shape and density of the powder also have a controlling effect on the
particle velocity. The fine particles accelerate and then decelerate very rapidly. The
coarse particles generally do not attain as high velocity as the fines do. The higher

the particle velocity the higher the coating density, provided that the particle has suffcient time to reach a molten state prior to impact. In order to take advantage of the high initial velocity of the fine powder size, shorter spray distances are used. A profile of average particle velocity with spray distance is shown in Fig. 16.

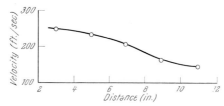

Fig. 16. Variation of Average Particle Velocity with Torch-to-Substrate Distance
(~50 μ particles) (From GRISAFFE and SPITZIG [18])

Powder particle size will vary according to the particular situation but generally it will fall within the range of 30 to 90 microns. The tendency is generally to use the fine particle size when ever possible, due to the higher velocity and shorter melting time requirements.

The *thermal conditioning* of the spray material is also affected by the powder feed rate. It is important to properly balance the powder feed rate and the arc gas enthalpy to obtain maximum particle melting. The powder feed rate is controlled by the carrier gas flow rate and the feed rate of the particular feed system. In the screw- or auger-type system the feed rate is controlled by the pitch and rotation speed of the screw. In the aspirator-type system the carrier-gas flow rate and orifice size control the feed rate. Greater control is generally obtained with the screw-feed system. An excessively high powder feed rate results in a high proportion of unmelted particles. Too low a feed rate is inefficient and can result in particle ablation.

The physical nature and purity of the spray powder can be controlled by the type of powder process employed and by the various prespraying treatments. In addition to high purity and close particle size control, irregular particles with low porosity are desired. Needlelike and platelike particles are undesirable because they are difficult to feed and to spray. Particles having a high porosity are also undesirable because the trapped gases may be retained in the deposited coating. Absorbed moisture is undesirable because it raises the thermal energy requirements and also causes powder agglomeration which hinders the feed process. Contamination on the surface of the powder can also be detrimental because it leads to areas of poor bonding in the deposited coating. However, in some cases an oxide or nitride film may act as a cementing phase between the coating particles. Elimination of moisture and other contaminants is generally accomplished by various calcining treatments and by maintaining proper powder storage facilities (desiccation, etc.).

Another important parameter of the spray process which must be controlled is the *spray environment*. Spray environments vary from air at standard conditions to a closely regulated atmosphere in a control chamber. The plasma effluent acts as an aspirator, drawing in gas from the surrounding environment and special precautions must be taken for spray powders which are subject to detrimental oxidation or nitride formation. This is also true for substrate materials which are subject to oxidation or nitriding. Various kinds of shrouds have been used to inhibit reactions with the environment. Both metal shrouds, to prevent aspiration, and inert cover gases have been reported. The addition of a small percentage of hydrogen gas in the effluent can be very effective because the hydrogen will combine with oxygen in the atmosphere, forming a protective sheath around the effluent. Gas jet rings around the exit nozzle are used to provide a blanketing cover gas around the effluent. Jets of inert cooling gas are also used to prevent substrate overheating and subsequent oxidation or warpage. However, as previously mentioned, the formation of oxide or nitride films on the spray particle or the substrate is not always detrimental. Often an oxide film will promote bonding between particles and between particles and the substrate. Other gases can also be added to the spray environment to promote or inhibit reactions in the effluent. For example carbonaceous gases are sometimes added to prevent decarburization of carbide powders during spraying [9, 15, 19].

For a completely regulated environment, atmosphere control chambers have been employed. Generally the chamber-sprayed coatings obtained are reported to be of higher quality than those obtained by conventional spraying techniques. In addition, materials which are easily oxidized or highly toxic usually have to be sprayed in control chambers. Higher temperature conditions tend to prevail in controlled-atmosphere chamber spraying. This may be by design, because the inhibiting effects of substrate oxidation are not present, or because of the difficulty in achieving adequate cooling. The ability to control spraying parameters more closely in chamber spraying results in higher deposition efficiencies (above 85%) and higher density coatings.

Although chamber spraying offers a number of advantages it also has it disadvantages. Major disadvantages are the higher spray costs (increased set-up time, higher gas consumption and limited operating time) and high initial cost for equipment. Other disadvantages are "fog" formation due to vapor condensation, which obscures viewing the operation, and "berry" formation in the nozzle of the torch, which limits operating time.

As pointed out in the above discussions, the number of process parameters, exclusive of those associated with the substrate, is quite large and somewhat foreboding at first evaluation. However, due to extensive research by equipment manufacturers and users, considerable control over the coating process is possible. Equipment manufacturers generally try to optimize equipment design and limit the number of variable process parameters. In addition most manufacturers supply the user with the average settings, the type of arc gas and the powder size to be used for spraying materials of interest. Using the manufacturer's data as a guide, precise settings can be determined by experimentation. Optimized settings for the spray equipment (arc-gas flow rates, powder feed rate, power levels, spray distance, gun angle and traverse rate) as well as powder size can be determined by the glass-slide test.

In the *glass-slide sampling test*, a glass slide is passed through the effluent at different distances from the exit nozzle and for different spraying conditions. The particles in the effluent are captured on the slide and studied with the aid of a microscope. A typical example of a properly melted ceramic particle on a glass

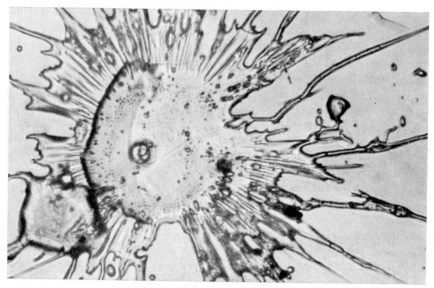

Fig. 17. Mullite Particle (2000×)

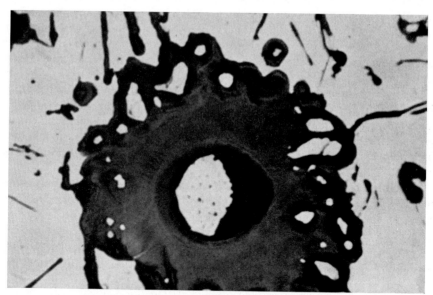

Fig. 18. Partially Melted Molybdenum Particle (2000×)

slide is shown in Fig. 17. This test is particularly useful for rapidly determining
the best spray distances. The effects of spray distance on the condition of a metal
particle in the plasma effluent are shown in Figs. 18, 19 and 20. In Fig. 18 the metal
particles have not become completely molten and the solid center core rebounded
from the glass slide. In Fig. 19 the metal particle is completely melted. Fig. 20
shows two metal particles which have begun to resolidify in the effluent. In

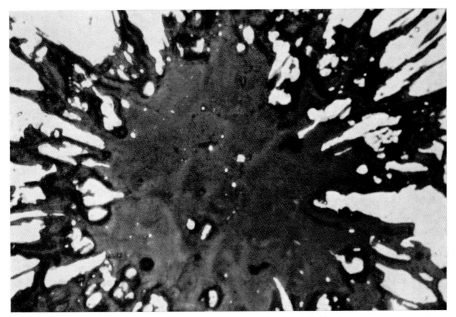

Fig. 19. Melted Molybdenum Particle (2000×)

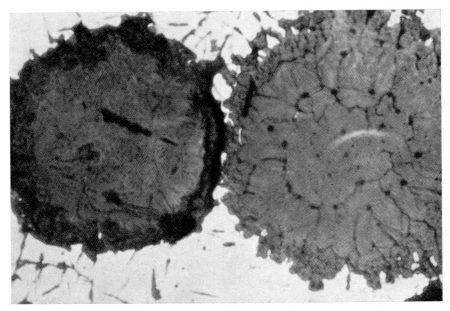

Fig. 20. Recooled Molybdenum Particle (2000×)

addition to the glass-slide test, the usual trial-and-error techniques are employed in determining the best spray conditions. Final spray settings are established by spraying and analyzing samples of the particular coating material on the substrate of interest.

   Although most research investigations have been of an empirical nature, there
have been a number of studies concerned with the fundamental reactions of
powder particles in the effluent and impacting upon a substrate. Of considerable
concern is the mechanism by which the particle is heated and propelled through
the effluent. It is believed that several mechanisms contribute to *particle heating*.
Thermal transfer by conduction through the boundary layer appears to be the
dominant mode of heating. Radiation, electron bombardment, and atom recombination on the surface of the powder, all contribute to the heating process. Radiation effects are not significant, and atom recombination is only a minor factor
which depends on the nature of the materials [20]. Metal surfaces have been found
to be better catalysts than glass or ceramic surfaces for atom recombination (see
chapter 7).

   When the powder particles enter the effluent they have zero axial velocity but
are subsequently accelerated by viscous drag. Heat is convectively transferred to
the boundary gas layer surrounding each particle as the effluent sweeps past the
particles. Gas velocity is reported at approximately 2,000 ft. per sec, as opposed
to a maximum particle velocity of 600 ft. per sec. Assuming boundary-layer conduction as the dominant heating mechanism, a model for determining heat transfer
rates to the powder particles was developed [20, 21] based on the assumption that
heating occurred in a finite length of effluent in which enthalpy and velocity were
uniform. Temperature gradients in the particle were not considered. It was further
postulated that, in the time required for the particle to traverse the specific
distance, it would become completely melted. The resulting equation is expressed
as follows:

$$\left[\frac{S(\lambda \Delta T)^2}{V\eta}\right] \geq \left[\frac{L^2 D^2}{16\varrho}\right] \tag{1}$$

where:

S distance the particle travels
λ mean boundary layer thermal conductivity
ΔT mean boundary layer temperature gradient
V mean plasma velocity
η mean plasma viscosity
L particle heat content/unit volume of liquid at its melting point
D mean particle diameter
ϱ particle density

   The left hand side of the equation is proportional to the enthalpy of the
effluent, approaching second power dependence. The enthalpy is also proportional
to the square root of the right hand side of the equation, $\left(\frac{LD}{V\sqrt{\varrho}}\right)$. The parameter
$\frac{L}{V\sqrt{\varrho}}$ can be considered as a measure of particle refractoriness [20]. This parameter can be very useful for determining the particle size of the spray powder or for
determining the particle size ratio for a blend of powders to be sprayed simultaneously [22, 23].

Deviation of experimental results from those theorized can be attributed to the fact that the model as developed does not account for the temperature, enthalpy and velocity gradients in the effluent. However, the velocity gradient acts in opposition to the enthalpy gradient and minimizes the effects of these factors. Correlation between the model and actual conditions is also complicated by the fact that not all of the particles can be injected into the center core of the effluent.

Heat transfer characteristics of various materials in a plasma flame have been determined using a specially designed probe [24]. For a 15-kilowatt arc using an argon/5% hydrogen gas mixture, the following rates for heat transfer were recorded:

copper	0.07 kcal per cm.^{2} sec
columbium	0.05 kcal per cm.^{2} sec
zirconium	0.05 kcal per cm.^{2} sec
molybdenum	0.04 kcal per cm.^{2} sec

These rates of heat transfer were increased 20 to 30% by increasing the hydrogen content by 5%. Using the heat transfer rates measured, the maximum particle size which could be completely melted in a residence time of 500 μsec was calculated. These values are shown in Table 2.

Table 2. *Maximum Particle Diameter for a Time of Residence of 500 μ sec. Restriction Imposed:*
$T_{center}/T_{surface} = 95\%$
(From SMITH [24])

Material	Maximum Particle Diameter (microns)	Value of $k/\varrho C_p$ (cm.^{2}/sec)
Copper	80	0.985
Zirconium	80	0.941
Tungsten	80	0.779
Molybdenum	80	0.468
Niobium	80	0.229
Alumina	59	0.0076
Spinel	44	0.0040
Zirconia	32	0.0025

Another model used to calculate heat transfer assumes that the surface of a particle with a radius R is instantaneously brought to a constant temperature T_s and for the duration of the dwell time, the interior of the particle is raised to the temperature of the surface. This model is expressed by the following equation [25]:

$$\frac{T}{T_s} = \left[1 + \frac{2R}{\pi r} \sum_{n=1}^{\infty} \frac{(-1)^n}{n} \sin \frac{n \pi r}{R} \exp. - \left(\frac{\alpha n \pi}{R} \right)^2 t \right] \tag{2}$$

where:
T temperature of particle at radius r
α thermal diffusivity $k/\varrho C_p$
t residence time
k thermal conductivity
C_p specific heat
ϱ density

The melting parameter derived from this equation for determining the optimum spray powder size is primarily a function of the thermal diffusivity of the spray material.

In traversing the distance between the spray gun and the substrate, the powder particles are subject to a number of chemical and physical interactions in addition to the normal heating process previously described. Gas adsorption, particle oxidation and nitride formation are examples of some of the more commonly observed particle-effluent reactions. These interactions generally are detrimental to the formation of a good coating. In addition to a change of solid to liquid state the powder particles may also undergo phase change, dissociation or volatilization. The condition of the powder in the effluent is very important because of the rapid quenching rate on substrate impact. It is not uncommon for the high-temperature phase to be retained in the coating (TiO_2 to $TiO_{(1-x)}$; α-Al_2O_3 to η- $+$ α-Al_2O_3; monoclinic ZrO_2 to tetragonal ZrO_2; zircon to glass $+$ ZrO_2). The condition of the particle on impact also has an important effect on the nature of the particle-substrate interactions [7, 26, 27].

Particle-substrate interactions are extremely important since this is the key phase of the coating process. To a large extent the nature of the particle-substrate interaction depends on the specific materials and their physical condition, and all encompassing generalizations are not possible.

Particle-substrate interaction begins with particle impact. Molten particles striking a smooth surface undergo severe deformation and flow radially outward from the point of impact. On rough surfaces there is very little atomization and this is a reduction in the degree of radial flow. Not all the impacting particles stick to the substrate. The fraction that adhere depends on the geometry and thermal condition of the substrate, as well as on the condition of the particle (size, velocity, temperature, viscosity, etc.).

The quenching rate of the impacted particles is generally quite rapid. A molten particle of Al_2O_3 0.1 mm. thick will remain molten for about 0.001 sec on a 1 mm. thick substrate of stainless steel, and about 0.03 sec on a glass substrate [26]. The rate of particle cooling has a considerable effect on particle-substrate and particle-particle interactions as well as on intraparticle reactions. To a large extent the particle cooling rate is determined by the substrate temperature, however the conductivity and specific heat of the substrate are also important factors to be considered [18]. As the first layer of particles is being quenched, the second layer of particles is striking the new substrate surface (the initial layer of particles). This results in a nonequilibrium type of process, since heat is drawn away from the inner face of the initial particle layer by the substrate as heat is added to the outer face by the new layer of impact particles (and by the impinging plasma gas).

Intraparticle reactions primarily affected by the quench rate are: the degree of particle fluidity, the resultant crystal phase, crystallite size and the degree of internal porosity. The normally rapid quench rate inhibits particle flow, retains the high temperature phase as previously described and inhibits crystallite growth and the degree of crystalline order. Crystallites forming within an impacting particle are very small and tend to be partially disordered. This is readily observed in the line broadening of X-ray diffraction patterns [7, 8].

The rapid quench rate also inhibits the particle-particle and particle-substrate interaction, thereby reducing the cohesive and adhesive strength of the coating. The reduced fluidity caused by rapid quenching is particularly detrimental to particle-substrate interactions. It has been shown that maintaining the substrate at an elevated temperature during the spray process does improve coating adherence [28, 29, 30], provided that the heating is not detrimental to the substrate.

The effect of using a heated mandrel surface and then maintaining the sprayed layers above 1000° C until the entire sample can be slow cooled has been studied for fabricating dense Al_2O_3 [31]. In this work cylindrical samples were sprayed on substrates of woven silica cloth or wire gauze. The silica cloth performed better than the wire gauze because of the lower degree of substrate distortion. To prevent cooling between sprayed layers, the traverse rate was carefully controlled, and for the larger samples a multiple system of spray guns simultaneously spraying overlapping zones was used. As sections were coated they were fed into a furnace where cooling below 1000° C was controlled at a rate of 70° C per hr. Samples prepared in this manner had high densities (1.5% open porosity for flat samples and 3 to 6% open porosity for cylindrical samples) and relatively high strength (MOR 20 to 26 × 10^3 psi).

The standard layered microstructure was prevalent on a macroscale; however, a considerable degree of columnar structure was observed and the deposited phase was α-Al_2O_3.

The *adherence* of the coating to the substrate is of major concern and most particle-substrate interactions are viewed in this perspective. The adherence (bonding) mechanisms operative between coating and substrate can be classified into three categories: mechanical, physical and chemical. The molten particles striking a roughened substrate surface conform to the surface topography. The mechanical interlocking between the coating and protrusions in the substrate surface is termed mechanical adherence. Substrate-coating adherence by VAN DER WAALS forces is classified as physical bonding. The formation of an interdiffusion zone or of an intermediate compound between the substrate and coating is generally termed chemical or metallurgical bonding. The adherence between a coating and substrate is generally a composite effect of two or more bonding mechanisms. The specific mechanism operative between a coating and substrate depends primarily on the materials used and the physical condition of these materials on impact.

The degree of mechanical interlocking varies with the materials involved, the spraying conditions and the prespray treatment of the substrate. Surface topography is considered a major factor in mechanical adherence. Various methods have been developed for improving the interlocking between substrate and coating. The most commonly used technique is surface abrasion by grit blasting. Sand, chilled iron grit, alumina, and silicon carbide are the major abrasives used. Alumina is the most commonly used abrasive because it tends to produce less warpage than chilled iron grit and less substrate contamination than silicon carbide [32]. The degree of abrasion is usually controlled by grit size and carrier-gas pressure. Re-use of grit is limited because of particle fracture.

In addition to grit blasting, surface grooving and undercutting (50 to 30°) are also used. However, these techniques are more common where large substrates and thick coatings are employed. Threading is frequently used for cylindrical substrates. Thorough surface cleaning (wirebrush) and degreasing is common to all surface preparation methods [9].

Various etching techniques have also been used to obtain a controlled substrate topography [33]. One technique of interest is the etching of the substrate surface to obtain single crystal protuberances. These single-crystal anchor points, having greater tensile strengths than the polycrystalline protuberances, should improve the mechanical adherence [34]. However, only limited data are supplied to support this concept.

Although the literature contains a number of reports correlating surface roughness with mechanical adherence it is difficult to justify a direct relationship because of the many interactions with abraded substrates [34, 35]. In addition to increasing the surface protuberances, the various abrading processes increase the surface area by exposing more clean, fresh metal for bonding [9]. Furthermore, these processes "work" the surface, thereby increasing the defect structure and making it more reactive toward the impacting particle [32, 29]. The reduction of particle fluidity by the roughened surface has a tendency to localize shrinkage strains and this also results in improved bonding strengths [9].

One technique commonly employed for improving the bond strength of a coating-substrate system is the application of a thin coating of a "self-bonding" or "bond coat" material between the substrate and the final coating layer. Molybdenum is one of the most frequently used materials for improving coating bond strength in this manner. Nickel aluminide, niobium, tantalum, nichrome and steel have also been used successfully in similar capacities [36, 22, 9]. Most of these materials have extremely high bond strengths even when the substrate is not abraded.

Because of their strong adherence, these materials can be used as an intermediate layer to promote high bond strength with thin or fragile substrates which must not be abraded. By using a coarse grained powder the intermediate layer can be made to have a very rough and irregular surface, which may further improve adherence of the top layer [36, 26, 37].

Various mechanisms have been suggested to explain the nature of adherence between these self-bonding materials and the substrate. Intermediate phase or alloy formation is the most frequently proposed mechanism. Two types of reactions are proposed for the formation of the intermediate phase:

1. Localized diffusion between the substrate and impacting particle [29, 38].
2. Localized fusion of the substrate and interaction between the molten particle and the melted substrate [39, 40].

The heat for the interaction is supplied by the impacting molten particle and the extent of interaction is controlled by the rate of particle quenching. From theoretical calculations it can be shown that sufficient heat and time are available for either localized diffusion or substrate fusion [29, 38].

Another mechanism frequently suggested for adherence is van der Waals forces or secondary valence bonds [9, 29]. Although van der Waals forces are normally thought of as weak forces, calculations have shown that the available attractive forces are at least a factor of ten greater than the measured strengths (30,000 to

50,000 psi calculated versus 1,500 to 5,000 psi measured) [29, 41]. Epitaxial effects and electrostatic attraction, due to electron transfer because of dissimilar work functions between the substrate and coating, must also be considered.

Concurrent with their high bond strength to the substrate "self-bonding" materials are particularly well suited for receiving sprayed layers of other materials. The exact reasons for the high bond strength observed between these intermediate layers and the oncoming layer of new coating materials are not completely known.

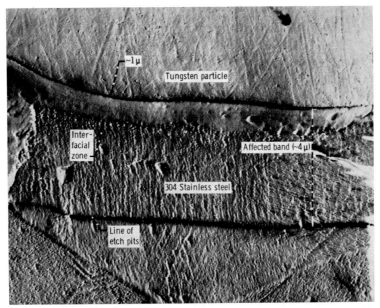

Fig. 21. Electron Micrograph of Cross-Section of Coating and Substrate for 3-inch Torch-to-Substrate Distance. Etchant, oxalic acid (9600×) (From SPITZIG and GRISAFFE [38])

It is believed that many of the same mechanisms operative at the substrate interface are responsible for the adherence between the intermediate zone and the cover coat (localized fusion and diffusion to form an intermediate phase, van der Waals forces, epitaxy, etc.).

Although a wide variety of metals, ceramics and plastics have been sprayed, only a very limited number have been analyzed for bonding mechanisms. The most frequently studied material is molybdenum. It is generally agreed that when molybdenum is sprayed on steel substrates it forms an intermediate phase, Fe_7Mo_6, at the interface. Both localized diffusion and substrate fusion have been proposed as the mechanism for this intermediate phase formation [29, 40]. Similar intermediate phase formation has been observed when molybdenum is sprayed on nichrome and cobalt [40]. This was not found to be the case when molybdenum was sprayed on chrome, and adherence was found to be quite poor. As an intermediate layer, molybdenum is well suited for receiving sprayed layers of other materials. The exact mechanisms responsible for this good adherence are not known. However, it would seem that the high surface energy of molybdenum would play an important role in any reactions at the interface.

Nickel aluminide (NiAl) coatings are prepared in a somewhat special manner in that the spray powder is a composite of spherical aluminum particles coated with a layer of nickel. The composition by weight is 80 Ni—20 Al. As the spray powder is heated in the effluent it is converted to NiAl. The conversion reaction is exothermic, and this excess heat is believed to cause localized fusion of the substrate. The molten substrate layer and the molten particle react at the interface to form an intermediate phase. This reaction has been observed for NiAl on steel substrates [39].

The adherence of tungsten coatings on steel, molybdenum and tungsten substrates have also been studied. On steel substrates the adherence appears to be due to localized diffusion and formation of an intermediate alloy phase. This is

Fig. 22. Cross-Sectional View of Coating-Substrate Interface. Etchant, Murakami's Reagent (5000×) (From Spitzig and Grisaffe [42])

illustrated in Fig. 21. A heat transfer analysis has shown that a tungsten particle at 3500° F can sufficiently raise the temperature of the steel substrate, directly beneath the particle, to accelerate diffusion rates. The depth of this effect has been calculated to be 1.3 microns. For tungsten on molybdenum or tungsten substrates, adherence appears to be due to an overgrowth of the boundary layer with recrystallized areas of the substrate [29, 38, 42], as shown in Fig. 22. This epitaxial effect is quite commonly observed when the spray powder and substrate are the same material or have reasonably close crystal structures and parameters (steel on steel, zinc on zinc, etc.).

Adherence of aluminum on steel is reported to be by localized diffusion and the formation of an aluminum-iron alloy phase at the interface [29]. The adherence mechanisms for Al_2O_3 on steel are still somewhat in question. Both intermediate phase formation and mechanical adherence have been reported [29, 35]. Localized diffusion and formation of $FeAl_2O_4$ at the interface is one of the mechanisms proposed for Al_2O_3 adherence [43]. Mechanical and physical adherence have been reported for copper and $BaTiO_3$ on steel [29, 44]. The same is generally reported for zirconium oxide; however metallurgical bonding has been reported for zirconium oxide on heated Inconel substrates [30].

Particle-particle bonding depends on many of the same mechanisms operative in particle-substrate adhesion. Here again, the nature of the reactions is greatly dependent on the characteristics of the particles (quench rate, viscosity, heat content, etc.). The dominant mechanisms appear to be diffusion, epitaxy and van der Waals forces. In the case of metals, the oxide or nitride film formed during spraying may act as a cementation phase between particles [9]. The effectiveness of this cementing phase depends to a large extent on thermal expansion match between the metal and its oxide or nitride phase. Contraction of the metal on cooling is generally greater than that of its oxides. The linear coefficients of expansion for several metals and their oxides are compared in Table 3.

Table 3. *Linear Coefficients of Expansion of Several Metals and Their Oxides*
(From MATTING and STEFFENS [29])

System	Metal		Oxide	
	Temperature Range (°C)	Coefficient (°C)$^{-1}$	Temperature Range (°C)	Coefficient (°C)$^{-1}$
Fe/FeO	0—900	$15.3 \cdot 10^{-6}$	100—1000	$12.2 \cdot 10^{-6}$
Fe/Fe$_2$O$_3$	0—900	$15.3 \cdot 10^{-6}$	20—900	$14.9 \cdot 10^{-6}$
Cu/Cu$_2$O	0—800	$18.6 \cdot 10^{-6}$	20—750	$4.3 \cdot 10^{-6}$
Cu/CuO	0—800	$18.6 \cdot 10^{-6}$	20—600	$9.3 \cdot 10^{-6}$
Ni/NiO	0—1000	$17.6 \cdot 10^{-6}$	20—1000	$17.1 \cdot 10^{-6}$

Flame sprayed coatings will vary in density from 80 to 95% of X-ray density. The porosity of the coatings is normally continuous, which would indicate incomplete cohesion. The microstructure and nature of the coatings are in many ways analogous to powdered compacts which have not been allowed to sinter to completion (sintering being stopped during the second stage of the process). This incomplete cohesion is also reflected in the bulk properties of the sprayed samples (Al_2O_3 tensile strength of 3000 psi, elastic modulus of 6×10^6 psi). In order to increase particle-particle contact it is important to have maximum particle fluidity and impact velocity. Post-spray sintering is also used to improve the density after spraying.

In any discussion of coating adherence and cohesion two additional factors which should be mentioned are:

1. substrate-plasma effluent interactions,
2. stress effects.

Bombardment of the substrate by the free electrons and ions in the plasma results in localized heating as well as in some degree of surface activation. Another cause of localized substrate heating is the heat generated by gas recombination on the surface. As mentioned earlier the effectiveness of a surface as a catalyst for gas recombination is very much material-dependent. The extent of plasma-substrate interaction on both adhesion and cohesion has never really been studied and it is difficult to do more than mention that these effects are occurring. The same is true for the associated effect of static charge, building up on both the spray powder and the substrate [32].

Various types of stresses are introduced into the coating and the substrate during the spraying process. These stresses have a decided effect on both the cohesive and adhesive strength of the coating. The temperature gradient produced

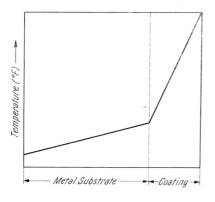

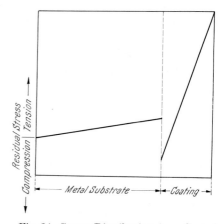

Fig. 23. Thermal Distribution through Coating and Substrate During Plasma Spraying (From MARYNOWSKI et al. [20])

Fig. 24. Stress Distribution in a Coated Sample After Cooling (From MARYNOWSKI et al. [20])

through the substrate and coating is probably the major cause of stress development. A typical thermal distribution is shown in Fig. 23. The resulting stress distribution on cooling is shown in Fig. 24. The tensile stress at the surface of the coating can cause cracking and spalling. The compressive stress at the interface tends to weaken the bond between the coating and substrate and may cause coating breakaway. For thin substrates these compressive stresses can cause severe deformation [9, 20, 26]. Various techniques to reduce thermal gradients during spraying have been developed, including cooling the back face of the substrate, gas cooling the front surface during spraying, or using a more rapid traversing speed [20, 45].

Other causes of stresses are the rapid rate of particle quenching, substrate warpage due to overheating, design features in the substrate (sharp corners, etc.)

and mismatch of thermal expansion coefficients between coating and substrate. One technique frequently employed to reduce the effects due to mismatched coefficients of expansion is the use of a coating layer between the substrate and

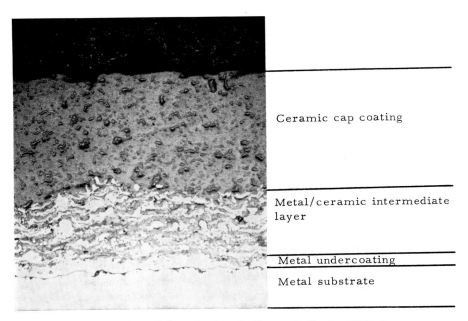

Ceramic cap coating

Metal/ceramic intermediate layer

Metal undercoating

Metal substrate

Fig. 25. Typical Cross-Section of a Composite Coating (250×)

top coating having an intermediate expansion coefficient. An example of this is shown in Fig. 25, where the intermediate layer is a mixture of the substrate and top coating materials.

After a coating is deposited, various treatments can be applied. The most common finishing treatment is grinding or polishing. Coating surfaces can be finished to an RMS of 2 to 4 microinches. For low-temperature applications, porous coatings can be impregnated and sealed with plastic resins or low melting glasses. High-density ZrO_2 films have been obtained by impregnating flame-sprayed discs with zirconyl nitrate solution and then firing to 1500° C [46]. These high-density films (5—10 mils thick) were translucent. Another technique for reducing coating porosity is to sinter the sprayed ware. Sintering is generally employed for thin-wall (no substrate) tungsten and molybdenum components [10, 47, 48, 49]. During post sintering the refractory metal will densify and recrystallize. The lamellar grain structure of the sprayed materials is converted to an equiaxial grain structure and the ductility is generally improved. One per cent nickel is sometimes added to increase the densification of the tungsten during sintering. The effect of reheating plasma sprayed Al_2O_3 has also been studied [27]. The primary interest of this study was the effect of reheating on the density and phase transformation (γ-Al_2O_3 to α-Al_2O_3). Reheating temperatures from 1150 to 1600° C were used. The inversion from γ-Al_2O_3 to α-Al_2O_3 results in an increased porosity of approximately 7% because of the increased bulk density of the new phase.

A technique for obtaining high-density alloy coatings is the plasma-fusion surfacing process. In this process powders which alloy with cobalt, nickel, chromium or iron are sprayed onto a molten substrate, alloying with the base metal during the solidification process. The deposition rates for this process are quite high, approximately twice that of ordinary welding techniques [50].

A number of different techniques are used in the preparation of thin-walled components. The most common method is to apply the coating on a substrate which can be removed easily after spraying (by etching, etc.). Two substrates frequently used for this purpose are aluminum and copper. Another method is to wet-spray a salt solution on a metal template and after the coating dries apply the flame spray coating on top of it. The coating can be easily removed by dissolving the salt in water [51].

3.3. Testing of Sprayed Coatings

Once a coating system is developed, it is important that it be well characterized so that its utilization can be maximized. A wide range of test techniques have been developed for evaluating coating systems. However, in the development or utilization of most tests it is important to recognize that it is a composite structure (coating and substrate) that is being tested. Both destructive and non-destructive techniques are available for evaluating the physical, chemical, thermal, mechanical, electrical and optical properties of coatings.

The adhesive and cohesive strength of a coating system is of major importance and a number of techniques have been developed to study these properties. These tests can generally be categorized in one of the following four groups:

(1) shear
(2) tensile
(3) bend
(4) impact

Shear bond strength is measured by applying a load parallel to the coating interface as shown in Fig. 26. The coated area is from $1/2$ to 1 inch in diameter

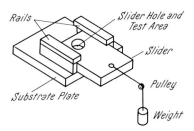

Fig. 26. Apparatus for Bond Shear-Strength Test (From GRISAFFE [35])

and 0.030 inch thick. The shear fixture can also be modified so that coating discs on both sides of the substrate can be tested simultaneously. The shear bond test is frequently used to test brittle coatings [35, 52].

In the *tensile bond test* a load is applied perpendicular to the coating interface. A coating 0.01 to 0.25 inch thick is applied to a circular or square substrate plate having an area of 1 to 3 square inches. The front face of the coating and the backface of the substrate are cemented to tensile fixtures with an epoxy adhesive. Only those specimens which fail at the metal-coating interface can be considered valid for bond strength measurement. Similar tensile tests on the coating itself (substrate removed) can be used to measure coating cohesion [19, 30, 52, 53, 54].

Another means of measuring coating adherence is the *bend test*. In this test the coated sample is bent until the coating begins to spall from the substrate. This test is also used to determine embrittlement effects in the substrate caused by the coating [55]. Various impact tests have also been developed for evaluating coating adherence and cohesion. One common type of test is the ballistic impact test where an air rifle is used to propel a steel pellet at the coating surface [55]. Generally this type of test is quite qualitative since it measures the amount of coating remaining after impact by recording the degree of electrical conductivity through the sample, or the amount of substrate surface area exposed.

Other equipment which can be used for studying adherence and cohesion are the optical and electron microscopes and the electron probe. The electron probe can be used to measure interdiffusion between substrate and coating. Adherence and cohesion can also be measured at elevated temperatures by modification of the test fixtures described above.

Discrepancies in reported adherence data are not uncommon in the literature. Different substrate preparations, spraying techniques and test procedures are the

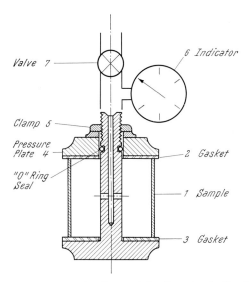

Fig. 27. Apparatus for Determining Porosity in Coatings
(From Ingham and Shepard [52])

main causes for differences in the data. Test samples are quite sensitive to preparation procedures as well as to sample size and shape. There is a general tendency for specimens cut from larger bulk samples to have lower coating-substrate adherence. Probably the major difficulty is the lack of a uniform set of test specifications for measuring coating adherence.

In addition to cohesion, a number of other properties can be measured on the coating layer itself. Of primary interest are elastic modulus, tensile strength, fatigue creep, and gas permeability [17, 19, 53].

Gas permeability is not only a measure of coating porosity but also a measure of the homogeneity and the degree of cohesion in the coating. A typical apparatus for permeability measurement is shown in Fig. 27.

A number of thermal, electrical and mechanical properties are measured on the composite system. Dielectric constants, electrical resistance and loss constants are measured by conventional methods.

Thermal conductivity is generally measured by the split-block technique and *thermal expansion* by dilatometry [44, 56]. In the measurement of these properties it is important to keep in mind the directional nature of the coating due to its lamellar structure. The effect of the substrate on the measured properties values must also be accounted for. This is particularly true for emittance measurements where substrate reflectance can effect the measured values if the coating is too thin (<0.016 inch) [23, 58, 60, 61].

Coatings are frequently applied in order to protect the substrate against deterioration by its environment. It, therefore, becomes important to characterize the chemical, thermal and mechanical resistance of the coating system. Low-temperature chemical and mechanical resistance is determined by standard methods. Various tests haxe been developed to determine the wear resistance of a coating [52, 55[.

Hardness tests (microhardness, Rockwell and scribe) are frequently employed to classify a coating system. These tests can be made at the coating surface, in an interior region of the coating, at the substrate-coating interface and in the substrate. Microhardness tests are frequently used to study the effects of the coating and the coating process on the substrate. Grit blasting (sand, silicon carbide, etc.), lapping, and grinding tests are used to measure abrasion resistance [9, 55, 57].

In addition to providing thermal insulation, coatings are frequently required to have a high resistance to abrasion, corrosion and thermal shock at elevated temperatures. *Abrasion resistance* to hot gases and hot particles is normally determined by exposing the surface of the coating to the effluent from a rocket motor exhaust or a plasma torch [22, 23, 62, 64]. Both plasma torches and rocket motor exhausts are also used to study the thermal-shock resistance of a coating system (see chapter 5). Normally a coated sample is cycled in and out of the hot effluent for a fixed period of time. Frequently the sample is cooled by a cold gas or water stream as it is removed from the hot effluent [22]. Various furnaces are also used for determining thermal shock resistances.

Oxidation is one of the major corrosion problems at high temperatures. Coatings are frequently employed for oxidation protection of the substrate. Both static and dynamic *oxidation tests* have been developed for coatings. In static tests, coating samples are held at 2000° to 3000° F for a predetermined period of time. Weight changes can be measured continuously or at specified intervals. Dynamic tests are conducted by passing air or steam over the coating surface. Oxidation resistance of coating systems at low pressures are also studied. Temperature stepdown

(30 minutes at 2500° F then, 30 minutes at 1800° F, then 30 minutes at 1400° F) tests are used to investigate "pest effects" in the coating [25, 55, 63].

Evaluation of a coating system without destroying the ware is extremely important, particularly in the production process. To meet these needs a number of *nondestructive tests* have been developed for coatings. The most commonly used technique is, of course, visual inspection or microscopic examination. To aid in the detection of surface defects a number of dye penetrants are also used (potassium ferricyanide, a suspension of fluorescent particles, etc.) [52, 55].

Radiographic techniques can be used to detect interior defects. By passing X-rays or gamma rays through a coating and substrate and then onto a photographic plate, defects can be detected due to changes in transmitted ray intensity. Complex shapes are difficult to analyze and defects must be greater than 2% of the total coating thickness in order to be detected [55].

Another technique for nondestructive testing which has recently received a good deal of attention is the use of ultrasonic waves for flaw detection. Ultrasonic waves introduced into a test sample will be modified due to interactions with defects. Longitudinal waves enter the test sample, travel in a straight line until they strike a boundary or discontinuity which reflects the wave back to a pickup system (transducer). The first boundary encountered is the surface of the coating, the second is the coating-substrate interface, and any signals received in the

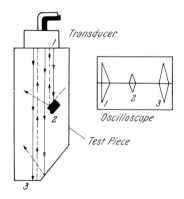

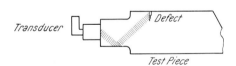

Fig. 28. Schematic Diagram Showing the Path of Longitudinal Ultrasonic Waves (From HUMINIK [55])

Fig. 29. Path of Shear Waves Used to Detect Defects Which Are Difficult to Locate with Longitudinal Waves (From HUMINIK [55])

interim can be attributed to defects in the interior (Fig. 28). Waves are sometimes introduced at an angle in order to pick up flaws not detectable by longitudinal waves (Fig. 29). If a very steep angle of entry is used the wave can be made to follow the surface contour (Fig. 30). This technique is very useful for detection of flaws at the surface [55].

In zirconium oxide coatings on Inconel, defects of $1/32$, $1/16$ and $1/8$ of an inch in length and from 300 to 500 µm. thick can be detected. However, defects at the interface cannot be detected because of the high reflection coefficient of the interface. Difficulties in flaw detection are also encountered due to scattering effects at grain boundaries, pores, surface roughness, etc. Complexity of sample shape can also be a problem in flaw detection [65, 66, 67].

A new nondestructive technique which is currently being evaluated for coating analysis is holographic interferometry. Initial results would indicate that this technique is very effective for determining residual stresses and detecting subsurface flaws in the coated samples.

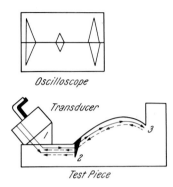

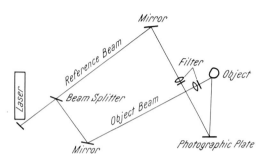

Fig. 30. Surface Waves Used to Detect Defects Close to the Surface (From HUMINIK [55])

Fig. 31. Experimental Arrangement for Preparing a Hologram

Holography is a two-stage laser photography process. In the first stage the image is recorded photographically using the experimental arrangement shown schematically in Fig. 31. The complex optical wave from the reflected object is superimposed on the reference wave front at the photographic plate. The optical

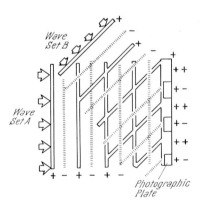

Fig. 32. Interaction of Reference Wave and Object Wave to Form a Hologram (From KOCK [109])

superposition of these coherent wave fronts forms an interference pattern (a series of Young's interference fringes) which is recorded on a high-resolution photographic emulsion as shown in Fig. 32. In the second stage of the process the object is recreated by illuminating the developed plate with the reference wave front.

The interference pattern on the plate acts as a diffraction grating which recreates the object wave front (as shown in Fig. 33). By looking through the hologram the viewer observes a true three-dimensional virtual image which is an exact replica of the object in size and position and displays both depth of field and parallax.

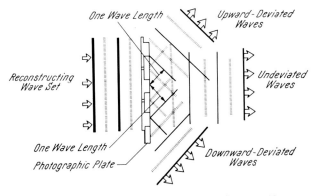

Fig. 33. Hologram Recreation (From Kock [109])

In holographic interferometry, the original holographic image is superimposed on the object (or on a hologram of the object) after it has been subjected to some form of stress (mechanical, thermal, etc.). Two different techniques are used to superimpose the stressed and unstressed conditions:

(1) time-lapse interferometry,
(2) real-time interferometry.

In the time-lapse technique (double-exposure method) two holograms are recorded on the same plate, one of the sample before it is stressed and one while it is under stress. In the reilluminated hologram, interference fringes are observed on the surface of the sample. These fringes are caused by the shift of surface points due to strain effects. In the real-time method, a hologram of the object is prepared in the conventional way and the processed plate is then replaced in its original position. The hologram is reilluminated with the reference beam and the real object (specimen) which is retained in the original position is illuminated as it was when the hologram was taken. If the recreated image is exactly superimposed on the undisturbed object no fringes will be observed. As soon as the real object is stressed fringes can be observed on the sample due to surface distortion.

The interference fringe pattern on the surface of the stressed sample is related to the surface deformation and reflects the mode of strain. If the fringes are formed on or near the surface of the sample (plane stress) the out-of-plane surface displacement (Δx) can be calculated from the relationship:

$$\Delta x = \frac{(2n+1)}{2} \lambda (1 + \sin \theta) \tag{3}$$

where:
n fringe order
λ wave length
θ the angle of incidence

If the fringes form at a considerable distance from the surface (several thousand wave lengths) due to complex sample distortion (rotation and translation), a much more complex mathematical relationship must be employed. Residual stresses, as well as defects at and below the surface of the sample will produce incongruities in the fringe patterns. An analysis of these incongruities can be conducted to determine the nature of the inhomogeneity [68, 69, 70, 109].

Several plasma-sprayed coatings of Al_2O_3 on 1-inch diameter disks and 1-inch square plates of $^1/_8$-inch steel, have been studied by holographic interferometry. A hologram of an Al_2O_3 sample loaded in compression, which does not

Fig. 34. Fringe Pattern for a Defect-Free Al_2O_3 Coating
(From HECHT [71])

appear to contain any residual stresses or flaws, is shown in Fig. 34. The slight inhomogeneity in the fringe pattern (nonuniform fringe spacing) is due to the fact that only one of the platens in the test device moved during loading. The fringe patterns obtained for coated samples containing residual stresses are shown in Figs. 35 and 36. Fig. 35 is a hologram of a disk sample showing the fringe lines at about 45° from the perpendicular. If the disk is rotated and loaded in different directions, the fringe line orientation will be altered. This effect is not observed for samples free of residual stresses. Fig. 36 shows a hologram of a square in which edge effects are quite pronounced.

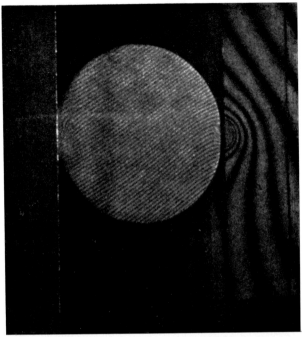

Fig. 35. Fringe Pattern for an Al₂O₃ Coating Containing Residual Stresses (From HECHT [71])

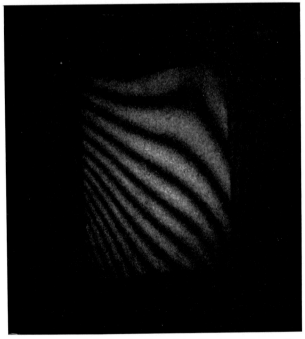

Fig. 36. Fringe Pattern for an Al₂O₃ Coating Containing Residual Stresses Due to Edge Effects
(From HECHT [71])

This same technique is also effective in detecting subsurface flaws, as shown in Fig. 37. In preparing this sample, a grease spot was placed on the steel substrate prior to spraying the Al_2O_3 coating. The complex fringe pattern indicates the presence of a debonding zone in the coating [71].

Fig. 37. Fringe Pattern for an Al_2O_3 Coating Containing a Defect at the Al_2O_3-Steel Interface (From HECHT [71])

Although, these holographic studies are preliminary, the initial results indicate that holographic interferometry can be an effective technique for detecting coating defects. However, a considerable amount of development is still required in order to correlate observed fringe patterns with specific defects and residual stresses.

3.4. Other Flame Spray Process

3.4.1. Combustion Spray Process

In the combustion spray process, oxygen and a fuel gas are ignited and the exhaust gas is used as the heat source for melting and for projecting the spray material. Fuel gases used in this process are acetylene, hydrogen, and propane. The flame temperature achieved by this process is approximately 5500° F. Material to be sprayed can be introduced into the hot flame in powder, rod or wire form.

A schematic diagram of a combustion spraying installation is shown in Fig. 38. The spray gun design for powder is shown in Fig. 39 and the design for rod spraying is shown in Fig. 40 [14, 51, 72].

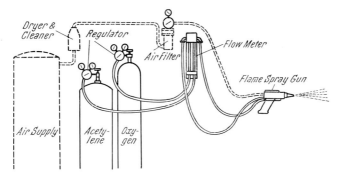

Fig. 38. Combustion-System Equipment (From WHEILDON [14])

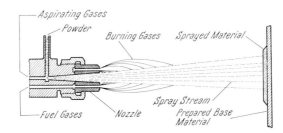

Fig. 39. The Powder Principle (From WHEILDON [14])

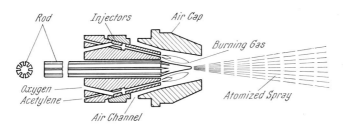

Fig. 40. Advanced Rod-Spray System (From WHEILDON [14])

In the powder spray system, material is introduced directly into a portion of the oxygen flow which aspirates it through the center of the burner nozzle where it is heated and carried to the substrate. Powders of almost all materials that have a liquidus can be sprayed. The average velocity of the sprayed powder will be between 100 and 200 ft. per sec [9].

The spray material can also be injected into the effluent in rod or wire form. This type of process has the advantage that the spray material cannot be propelled into the effluent until it is completely molten, as opposed to powder particles which may remain incompletely melted on impact. As the material is melted on the tip of the rod or wire it is atomized by a current of compressed air and propelled toward the substrate. The spray rate is controlled by the melting rate of the rod which is, in turn, a function of the gas enthalpy and the melting characteristics of the rod. The viscosity of the liquid phase is also an important factor. The average

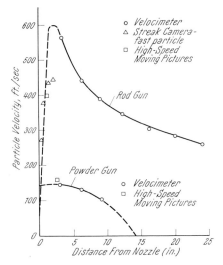

Fig. 41. Comparison of Average Particle Velocities (From MOORE *et al.* [73])

particle velocity in the rod spray process is 400 to 600 ft. per sec. This is somewhat higher then the velocity of the powder-sprayed particles as shown in Fig. 41 [26, 73].

Not all materials can be fabricated into wire or rod form. Most wire spraying is limited to the metals, while a number of ceramics have been fabricated in rod form (Al_2O_3, ZrO_2, Cr_2O_3, $ZrSiO_4$). The diameter of the rod or wire can be varied from $1/8$ to $1/4$ of an inch. The rods are generally fluted to increase the spray rate and decrease the susceptibility to thermal shock. The rod spray process is commonly termed the *Rokide* Spray Process [14, 51].

Coatings prepared by the combustion spray process will have a lamellar microstructure with continuous porosity ranging from 10 to 20%. Excess air and moisture in the exhaust gas can have a number of detrimental effects on the nature of the sprayed coating. Moisture collecting on the surface of the substrate or in a layer of the coating, inhibits good bonding by preventing intimate contact of the particles. The oxidation of the spray powder in the effluent by excess air or moisture can also be detrimental.

3.4.2. Detonation Process

In the very broadest sense the detonation or *Flame-Plating* process can also be considered as a combustion process. In this process a mixture of acetylene, oxygen and powdered coating material is pressure-fed into a chamber and detonated by a spark plug. The detonation wave heats and accelerates the particles to the substrate. A schematic representation of this process is shown in Fig. 42. The basic operating cycle is as follows:

"The oxygen (O_2) and acetylene (C_2H_2) valves open, allowing oxygen and acetylene to be injected into the barrel. Powder from the feeder is carried by nitrogen (N_2) gas under pressure to the barrel. When the barrel is nearly full of fuel gas and powder, the oxygen and acetylene valves close and the nitrogen valve opens purging the fuel gases from the valve chambers and passages leading to the barrel. Just before the nitrogen reaches the spark plug the magneto fires the plug. A detonation sweeps through the oxyacetylene charge in the barrel accelerating the particles of powder to a very high velocity, melting and depositing them on the substrate. The barrel is detonated approximately 4 times per second (240 times

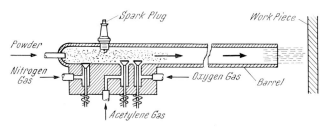

Fig. 42. Detonation Principle (From WHEILDON [14])

per minute). The detonated gas has a velocity of 9600 ft. per sec, a pressure of 6 atmospheres and a temperature of 9000° F. Powder particles in the detonation attain an average velocity of 2400 ft. per sec. The rate of deposition is about 0.25 mil over a one inch circular area for each detonation. The detonation equipment is operated automatically in a soundproof, explosion-proof room, because of the high sound level of 150 decibels and the danger of explosion" [51].

A number of ceramic and metal coatings have been prepared by the detonation process (Al_2O_3, Cr_2O_3, WC, WC + Co, CrC + NiCr, NiCr, NiFe, Cu + Ni + In, and Ni + Cr + BN). All of the coatings prepared have been reported to have a very high density (99%) and a very high bond strength (6000 to 10,000 psi) [51, 72, 9, 74, 75].

3.4.3. Liquid Fuel Gun

Another modification of the combustion spray process is the liquid fuel gun. In this process a liquid fuel, such as heavy fuel oil, and oxygen are used for combustion. The powder to be sprayed is mixed with the oil to give a suspension which results in a higher concentration and a more uniform distribution of powder in the flame. In addition this technique results in better heat transfer than normally achieved in the other spray processes. The spray gun is a water-cooled barrel 3 inches in diameter and 3 feet long. The gun was developed primarily for forming refractory linings in a furnace and therefore spraying is done on substrates which are relatively hot (1500° F). With the current design the powder deposition rate is 5 to 10 lbs. per min. There are also plans for a gun 8 inches in diameter by 5 feet long which is expected to spray 250 lbs. per min.

A number of oxide materials have been sprayed by this process (alumina, mullite and silica). The minimum porosity attained is 1% although porosities of 20%

Table 4. *Comparison of*
(After PALERMO

| Process | Plasma Arc | Combustion | | Liquid Fuel Gun |
		Oxyacetylene Unfused	Metallizing Fused (1)	
Choice of Coating Material	Metals, Ceramics and Plastics	Nonreactive metals; refractories with melting point 5000° F	"Self-Fluxing" alloys	Al_2O_3, Mullite silica, other oxides
Choice of Base Material	Almost all metals and ceramics; some organic materials	Almost all metals and ceramics	Metals with melting point 2000° F	Metals and ceramics
Normal Processing Temperature	Usually less than 250° F; up to 400° F for a few coatings	500—600° F	1850—2150° F	Substrate heated to 3000° F
Particle Velocity	600 ft./sec	150—400 ft./sec	150—400 ft./sec	1000 ft./sec
Type of Bond	Mechanical, physical, metallurgical	Mechanical, metallurgical, physical	Metallurgical	
Dimensional Limits on Base (3)	0.025 in. dia min. no max. limit	0.004 in. dia min. no max. limit	0.06 in. dia min. no max. limit	
Coating Thickness and Tolerances, in. (4)	0.002—0.1, ±0.001	0.005—0.2, ±0.003	0.005—0.2, ±0.005	Coatings reported to be several inches thick
Surface Finish rms microinch As Applied As Ground	75—125 10 typically, as low as 2—4	150—300 1—12	150—300 5	
Equipment Cost	$ 5,000—$ 8,000	$ 500	$ 1,500	

(1) As sprayed coating is heated to melting point to consolidate coating and produce metallurgical alloying of coating with base.
(2) Flame-Plating.

Flame Spray Processes
and POLTER [76])

Detonation (2)	Metallic Arc and Gas Welding	*Rokide*	Electric Arc
Tungsten Carbide with selected matrices, selected oxides	Ferrous and Cobalt base alloys; tungsten carbide-alloy, hard facing alloys	Selected ceramics, Al_2O_3, ZrO_2, Cr_2O_3, $ZrSiO_4$	Most metals
Almost all metals and ceramics	Weldable materials usually ferrous base	Most metals and some non-metals	Most metals
400° F	Locally from 1800° F up to melting point of base	4—650° F	
2400 ft./sec		400—600 ft./sec	600 ft./sec
Intimate mechanical	Metallurgical	Mechanical, physical, metallurgical	Metallurgical mechanical
0.2 in. dia min. to 60 in. max. dia	Very small to very large	Approx. 0.010″ D no max. limit	
0.001—0.012, ±0.001	0.1—0.5, variable	0.001—0.002	
125 as low as 1	Irregular As low as 5; often marred by blow holes	200—300 10—16	
			$ 3,000

(3) These are practical limits; in some cases special techniques permit application of coating to thinner sections.

(4) These are common values; limits often vary with base and coating materials.

can be obtained by modifying the spray parameters. The reported deposition efficiency is approximately 75% [7].

3.4.4. The Electric Arc Process

In this process a motor generator provides direct current to energize two consumable wire electrodes as shown in Fig. 43. When the wire feed is started, the tips make contact, strike an arc and melt. The molten metal is propelled toward

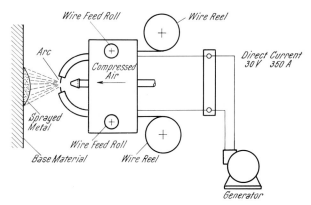

Fig. 43. The Electric-Arc Process (After METCO)

the substrate by a blast of compressed air. The average particle velocity is 600 ft. per sec. Coatings prepared by this method are reported to have high density, good bond strength and a lower degree of particle oxidation [16].

3.4.5. Comparison of Flame Spray Processes

A compilation of selected properties associated with each of the flame spray processes is presented in Table 4 for purposes of comparison. The combustion process is less expensive and requires less complex equipment than the plasma arc process. However, the combustion process is limited to materials which have melting points below about 5000° F, and which are not severely deteriorated by the oxidizing atmosphere. Coatings produced by the combustion process tend to have lower density, weaker bond strength and looser tolerances than those obtained by the plasma spray process.

Coatings prepared by the detonation process have higher density and greater bond strength than normally found for plasma-sprayed coatings. In addition the tolerances are closer and the surface finish and physical properties are somewhat better. However, the cost of the detonation process is greater and it is not effective for large surface area coverage or coverage of complex shapes [76, 77].

It is difficult to assess the effectiveness of the electric arc process at this time since its use has been limited. It is also difficult to assess the newly developed liquid-fuel gun at this time. However, from preliminary information it appears that coatings prepared by this process will have higher densities and can be sprayed to much greater thicknesses.

4. Materials and Applications for Plasma Spraying

Flame sprayed coatings have found wide utilization as protective and decorative coatings. In addition, this process has been effective for fabricating complex thin wall components. Numerous applications for metal, ceramic and organic coatings are described in the literature. A summary of the applications for plasma sprayed materials is presented in Table 5.

In addition to the formation of coatings and free-standing forms flame-spraying equipment can also be used for preparing ceramic and metal powders to be employed in other fabrication processes. The high temperature of the effluent and the large number of gas environments (inert, reducing, oxidizing, etc.) which can be achieved in the flame spray process enables the generation of many chemical and physical reactions not otherwise possible. In addition, the products formed at high temperature in the effluent can be retained at room temperature due to the rapid quenching possible with this process. With flame-spraying equipment, material preparation processes requiring thermal decomposition, vapor-gas phase reactions and liquid-phase reactions are possible.

Preparation of submicron carbon from suitable hydrocarbons, molybdenum from the dissociation of molybdenite (MoS_2), and Al_2O_3, BeO and SiO_2 from beryl are some of the examples, reported in the literature, where flame spraying equipment has been used for the preparation of materials by thermal decomposition. Various oxides, carbides and nitrides can also be prepared from vapor-gas reactions by properly controlling the composition of the effluent gas. Powder purification and the preparation of ultrafine particles of powder (submicron) are also feasible with flame-spraying equipment [78].

Another important application for the flame-spraying equipment is the production of spherical particles. If the molten particles are projected into a cooling chamber and allowed to solidify before impact, the resultant particles will be spheroidized. Particles ranging from a few microns to a few hundred microns in size can be obtained. Since it is difficult to produce the large diameter particles ($> \sim 150\mu$) a good deal of care must be exercised [78, 81]. For a number of ceramic powders this same technique also can be employed to retain the material in its glassy state. Glass spheres of $BaTiO_3$, Al_2O_3 and ZrO_2 have been obtained in this manner [79]. The rapid quenching possible with the flame spray process provides the only means for obtaining the glassy state for a number of ceramic compositions.

Ceramic flame-sprayed coatings include the oxides, carbides, borides, nitrides and silicides, as well as a number of glass compositions. These ceramic coatings have been used for thermal insulation, electrical insulation and for erosion and corrosion resistance. A compilation of reported properties for plasma-sprayed

Table 5. *Summary of Coating Application For Plasma Sprayed Materials*
(From FAIRLIE [77])

Material	Melting Point (°F)	Possible Areas of Usefulness
Alumina-50% Titania Blend	3490	Strongly adherent coating, temperature-resistant, moderate resistance to thermal shock (Maximum usable temperature 2000° F in oxidizing atmosphere).
Alumina-40% Nickel Blend	2650	High-temperature coatings where thermal shock is a problem.
Aluminum Oxide-Titania Trace	3500	Low-cost ceramic provides good temperature, wear and impact resistance. Low reaction to molten metals (Maximum usable temperature 2800° F oxidizing atmosphere).
Aluminum Oxide-Pure	3600	Good wear, heat and corrosion resistance, good insulator; stable in oxidizing or reducing atmosphere at 2800° F; resists product pickup on annealing-furnace rolls and has low reactivity to molten metals. Thermal-control coating.
Boron Carbide	4400	Cermet component, nuclear shielding and controlrod material; wear- and temperature-resistant.
Calcium Zirconate	4250	Thermal barrier, coatings. Resistant to wetting by various metals, and can be used for coating melting pots and allied equipment. Dense, hard, abrasion-resistant coatings with very good bond. Maximum usable temperature 2400° F in oxidizing atmosphere.
Ceric Oxide	4712	Thermal barriers, combustion catalyst.
Chromium	3435	Bearing surfaces at low to medium temperatures. Corrosion-resistant coatings when properly sealed.
Chromium Carbide	3700	Wear-resistant coatings, mixture with metal powders for cermet coatings and wear resistance at higher temperatures.
Chromium Carbide-40% Cobalt Blend	2700	Wear-resistant coatings, particularly on aluminum and other non-ferrous metals where bond is excellent.
Cobalt	2715	Dense, strongly adherent coatings. Mixture with ceramics for cermets.
Magnesium Zirconate	3830	Thermal-barrier coatings, resistant to wetting by various metals, and can be used for coating melting pots and allied equipment. Especially useful in lining graphite crucibles used in melting and refining uranium.
Molybdenum	4760	Bonding metal and ceramic coatings for low-temperature use, hard wear-resistant surfaces. Also for electrical contacts.
Nickel	2650	Corrosion-resistant coatings when properly sealed, mixed with ceramics to form cermet coatings.

Table 5 (*continued*)

Material	Melting Point (°F)	Possible Areas of Usefulness
Rare Earth Oxides	over 4000	Thermal barrier and combustion catalyst. Recommended for coating of diesel pistons and combustion chambers. Reduces fractures in combustion chamber.
Titanium Carbide	5600	Cermet component, high-temperature electrical conductor. Good thermal and oxidation resistance. (Maximum useful temperature 2500° F in oxidizing atmosphere).
Titanium Oxide	3490	Hard, abrasion-resistant, minimum-porosity coatings with excellent adhesion to base. For mixture with other ceramic and metal powders to improve physical properties of the coatings.
Tungsten	6170	Liners for rocket-engine throats and tail cones.
Tungsten Carbide	5200	Excellent blast-erosion qualities as sprayed for high-temperature applications i.e. rocket nozzle coatings. Excellent abrasion resistance through temperature range up to 1500° F.
Tungsten Carbide- 12% Cobalt Blend	Softens at above 2715	Good adherence with good wear and shock resistance.
Zirconia 40% Cobalt Blend	above 3000	High-temperature cermet coating. High hardness and abrasion resistance. Retards oxidation of base materials and has excellent thermal-shock resistance.
Zirconium Carbide	6400	Cermet component, refractory material, high-temperature electrical conductor.
Zirconium Oxide (Hafnia-Free Lime Stabilized)	4700	Thermal-barrier coatings for nuclear applications.
Zirconium Oxide	4700	Resistant to high-temperature oxidation to 4100° F, resists reaction to molten metals and pickup of product on annealing-furnace rolls and used on forging dies.
Zirconium Silicate	3000	Thermal-barrier coating with good resistance to wetting by molten metals up to 2000° F.

Many additional metals with their oxides, silicates, nitrides, borides and carbides are available for research or special applications. All materials which melt without chemical change can be sprayed with the plasma process.

ceramic coatings is presented in Table 6. Similar compilations are also presented for *Rokide* and *Flame-Plated* coatings in Tables 7 and 8, respectively.

The most frequently used oxide coatings are alumina and zirconia. Alumina is used primarily for thermal and electrical resistance [82, 83]. Flame-sprayed

Table 6. *Properties of Flame-Sprayed Materials*

Material	Density % of theoretical	Melting Point (°F)	Specific Heat (Btu/lb.°F)	Thermal Conductivity (Btu/hr./ft.²/in./°F)	Thermal Expansion coeff. (in./in./°F)	Tensile Strength (psi)	Compressive Strength (psi)	Strain %
Al_2O_3	85—91%	3600	0.3	12—20	4—$7 \cdot 10^{-6}$	3—$12 \cdot 10^3$	20—$35 \cdot 10^3$	0.7
$BaTiO_3$	80—90%							
Be	85—90%					$24 \cdot 10^3$	$40 \cdot 10^3$ $60 \cdot 10^3$ aft. sint	
ZrB_2	93%						$10 \cdot 10^3$	
CrC	96—97%						4—$10 \cdot 10^3$	
HfC	81—87%						32—$48 \cdot 10^3$	
TaC							32—$42 \cdot 10^3$	
ZrC							$16 \cdot 10^3$	
WC	90%							
CeO_2	92%							
Cr_2O_3	90—95%	3000	0.2	18	$5 \cdot 10^{-6}$		$30 \cdot 10^3$	
HfO_2	87%							
MgO	90—95%	3500	0.25	18	$4.5 \cdot 10^{-6}$			
Mo	88—93%	4760			$3.5 \cdot 10^{-6}$	4—$25 \cdot 10^3$ 50,000 aft. sint.		0.3
NiAl	85—95%					5—$6 \cdot 10^3$		
TiN							$37 \cdot 10^3$	
ZrN							$43 \cdot 10^3$	
TiO_2	70—85%	3200		15	3.9—$4.6 \cdot 10^{-6}$			
W	80—93% to 98% after sintering				$2.7 \cdot 10^{-6}$	10—$21 \cdot 10^3$ 4500 aft.sint.		
ZrO_2	85—95%	4500	0.17	4—8	3—$7 \cdot 10^{-6}$	1—$1.4 \cdot 10^3$		
$ZrSiO_4$	85—90%	3000	0.15	15	$4 \cdot 10^{-6}$		20—$30 \cdot 10^3$	1.4

Table 6 (continued)

Material	Modulus of Rupture (psi)	Modulus of Elasticity (psi)	Hardness	Adherence (psi)	Emittance	Dielectric Strength (V/mil.)	Dielectric Constant
Al$_2$O$_3$		6 · 10^6	1.4—2 · 10^3 knoop	1.5 · 10^3	.8 at 400° F .5 at 1400° F .4 at 1800° F	200	10
BaTiO$_3$		29 · 10^6		1035			
Be				2—3 · 10^3		90—300	23.5—500
ZrB$_2$			1,000 VHN				
CrC							
HfC			400—800 VHN				
TaC			300—1000 VHN				
ZrC			800 VHN				
WC			C-38				
CeO$_2$	8,670			3,000			
Cr$_2$O$_3$			1200 VHN		.9 (250° F) .75 (1800° F)		
HfO$_2$			250—310 VHN				
MgO							
Mo	35 · 10^3	50 · 10^6	400—500 knoop	2700			
NiAl			Rb 75	3—3.6 · 10^3			
TiN			781—1095 VHN				
ZrN			623—1120 VHN				
TiO$_2$			1500 knoop				
W	20,000 55—75 · 10^3 aft. sint.	40—60 · 10^6	A 50	1000			
ZrO$_2$	3,500		600—800 VHN	1500—1600	0.7—0.8 400—1200° F	50—100	35
ZrSiO$_2$			1000 knoop	1000—2000			15

Table 7a. *Physical Properties of Rokide Coatings*
(From WHEILDON [14])

Types of Coatings	Color	Crystal Form	Bulk Density (Average gm./cc)	Crystal Hardness Knoop Scale	Porosity (%)	Permeability
Rokide A Aluminum Oxide	White	Gamma Type	3.3	2000	8 (7% open)	Slight
Rokide Z Zirconium Oxide	Light Tan	Cubic	5.2	1000	8 (7% open)	Slight
Rokide ZS Zirconium Silicate	Light Tan	Cubic ZrO_2 in Siliceous glass	3.8	1000	8 (4% open)	Slight
Rokide C Chrome Oxide	Black	Hexagonal	4.6	1900	4 (2% open)	Very Slight
Rokide MA Magnesium Aluminate	White	Cubic	3.3		6 (4% open)	Slight
Magnesium Zirconate		Cubic	4.52		4.8 (4.5% open)	
Calcium Zirconate		Cubic	4.414		5.7 (4.5% open)	
Strontium Zirconate		Orthorhombic	4.71		9.8 (7.8% open)	
Barium Zirconate		Cubic	5.25		10.0 (5.1% open)	

Table 16. *Mechanical Properties of Rokide Coatings*
(From WHEILDON [14])

Types of Coatings	Compressive Strength psi at Ambient Temperature	Adherence to Steel psi (approx.)	Minimum Profilometer Surface Finish RMS Range (Microinches)	Abrasion Resistance Rubbing Wear Impact Abras.	Coefficient of Friction	Strain % Elongation per Unit of Lgth. (0.020″ Thk. Coat.)	Vibration Cycles
Rokide A Aluminum Oxide	37,000	Steel 1000 Non Ferrous 600	As Coated 200—300 As Ground 30—50 As Lapped 25—45	Very Good	.10 Against 440-C SS	.7	200,000,000 Cycles Without Failure (15,000—16,000 psi Cyclic Stress)
Rokide Z Zirconium Oxide	21,000	Steel 1,000 Non Ferrous 600	As Coated 200—300 As Ground 30—50 As Lapped 25—45	Good	.10 Against 440-C SS	1.4	
Rokide ZS Zirconium Silicate		Steel 1,000 Non Ferrous 600	As Coated 200—300 As Ground 30—50 As Lapped 25—45	Good	.10 Against 440-C SS	.7	
Rokide C Chrome Oxide	105,000		As Coated 200—300 As Ground 15—25 As Lapped 10—20	Very Good	.11 Against Brass	1.3	
Rokide MA Magnesium Aluminate			As Coated 200—300 As Ground 30—50 As Lapped 25—45	Good			
Magnesium Zirconate		1,000					
Calcium Zirconate		1,125					
Strontium Zirconate		1,129					
Barium Zirconate		1,153					

Table 7c. Electrical Properties of Rokide Coatings
(From WHELDON [14])

Types of Coatings	Conductivity	Electrical Resistivity Ohm-Inch	Dielectric Strength AC Volts Per Mil. Thk.	Dielectric Constant at 0—600° F	Dissipation Factor at 0—600° F	Loss Factor
Rokide A Aluminum Oxide	Nonconductor	$4.5 \cdot 10^6$ at 500° F $1.2 \cdot 10^5$ at 800° F	.005″ Thk.-160 .010″ Thk.-120 .020″ Thk. - 65 .030″ Thk. - 48 +	10	.05	.5
Rokide Z Zirconium Oxide	Nonconductor (Amb. Temp.) Increases Rapidly at 1200° C	$2.7 \cdot 10^4$ at 500° F $2.1 \cdot 10^2$ at 800° F		35	.15	5.25
Rokide ZS Zirconium Silicate	Nonconductor	$1.1 \cdot 10^6$ at 500° F $4 \cdot 10^2$ at 800° F		15	.08	1.20
Rokide C Chrome Oxide	Nonconductor					
Rokide MA Magnesium Aluminate	Nonconductor					
Magnesium Zirconate			150			
Calcium Zirconate			70			
Strontium Zirconate			130			
Barium Zirconate			140			

Types of Coatings	Melt Temp. (°F)	Mean Specific Heat (Btu/lb./°F)	Coefficient of Expansion (in./in./°F)	Conductivity (Btu/hr./ft.² in./°F) (Coating Only)	Total Emittance	Thermal shock Resistance	Chemical Composition (%)	Resistance to Acids and Alkalis
Rokide A Aluminum Oxide	3600	.28 (90—3100° F)	4.1·10⁻⁶ (70—2250° F)	19 (1000—2000° F)	.8—.4 (200—1000° C)	Good	Pure Aluminum Oxide 98.6 Al_2O_3	Good Good* (Except Hot)
Rokide Z Zirconium Oxide	4500	.175 (80—2550° F)	5.4·10⁻⁶ (70—2250° F)	8 (1000—2000° F)	.7—.3 (200—1000° C)	Very Good	ZrO_2 +2% HfO_2 +3.5—5% CaO	Good Good* (Except HF)
Rokide ZS Zirconium Silicate	3000	.15 Est	4.2·10⁻⁶ (70°—1100° F)	15 (1000—2000° F)	.7—.3 (200—1000° C)	Good	65 ZrO_2 34 SiO_2	Fair Good* (Except HF) (Except Hot)
Rokide C Chrome Oxide	3000	.2 (60 —2700° F)	5.0·10⁻⁶ (70—2000° F)	18 (Est) (1000—2000° F)	.8—.9 (100—1200° C)	Moderate	85 Cr_2O_3	Good Good* (Except HF) (Except Hot)
Rokide MA Magnesium Aluminum	3500	.25 (70—1832° F)	4.5·10⁻⁶ (70—2000° F)	18	.7—.3 (100—1200° C)	Good	98 $MgO\text{-}Al_2O_3$	Good Good* (Except HF) (Except Hot)
Magnesium Zirconate	3880		4.0·10⁻⁶ (70—2012° F)					
Calcium Zirconate	4245		4.6·10⁻⁶ (70—2012° F)					
Strontium Zirconate	5070		4.7·10⁻⁶ (70—2012° F)					
Barium Zirconate	4870		3.5·10⁻⁶ (70—2012° F)					
* Alkalis								

Table 8. *Properties and Characteristics of Flame-Plated Coatings (Detonation Process)*
(From McGeary and Koffskey [81])

Properties	Coating Composition (wt. %) and Designation	
	WC+9% Co (LW-1; AMS 2435)	WC+13% Co (LW-1 N 30)
Vickers hardness (300 kg. load)	1300	1150
Maximum temperature in oxidizing atmosphere (F)	1000	1000
Coefficient of thermal expansion (in./in. °F)	$4.5 \cdot 10^6$ (70—1000° F)	
Modulus of rupture (psi)	80,000	90,000
Modulus of elasticity (psi)	$31 \cdot 10^6$	$31 \cdot 10^6$
Porosity (%)	1	1
Specific Gravity	14.2	13.2
Thermal Conductivity (Btu/ft./hr./°F sq. ft.) at 200° F at 500° F	3.7 5.3	3.7 5.3
Characteristics of Flame-Plated Coatings	Extreme wear resistance	Excellent wear resistance; increased resistance to mechanical and thermal shock.

Table 8 (*continued*)

Properties	Coating Composition (wt. %) and Designation	
	WC+15% Co (LW-1 N 40)	60% Al_2O_3 +40% TiO_3 (LA-7)
Vickers Hardness (300 kg. load)	1050	950
Maximum temperature in oxidizing atmosphere (F)	1000	1300
Coefficient of thermal expansion (in./in.°F)	$4.7 \cdot 10^6$ (70—1000 °F)	
Modulus of rupture (psi)	100,000	19,000
Modulus of elasticity (psi)	$31 \cdot 10^6$	$11 \cdot 10^6$
Porosity (%)	1	1
Specific Gravity		
Thermal Conductivity (Btu/ft./hr./°F sq. ft.) at 200° F at 500° F	3.7 5.3	
Characteristics of Flame-Plated Coatings	Greatest resistance to mechanical and thermal shock. Excellent wear resistance.	Coating is a true semiconductor. Good wear resistance and mating properties. Excellent textile material.

Table 8 *(continued)*

Properties	Coating Composition (wt. %) and Designation	
	25% WC+5% Ni+ mixed W-Cr carbides (LW-5)	80% Cr_3C_2+ 20% Nichrome (LC-1B)
Vickers hardness (300 kg. load)	1075	700
Maximum temperature in oxidizing atmosphere (F)	1400	1800
Coefficient of thermal expansion (in./in.°F)	$4.6 \cdot 10^6$ (70—1400° F)	$6.4 \cdot 10^6$ (70—1800° F)
Modulus of rupture (psi)	40,000	70,000
Modulus of elasticity (psi)	$17 \cdot 10^6$	$18 \cdot 10^6$
Porosity (%)	1	1
Specific Gravity	10.1	6.5
Thermal Conductivity (Btu/ft./hr./°F sq. ft.)		
at 200° F	3.1	
at 500° F	2.8	
Characteristics of Flame-Plated Coatings	Resistance to wear at higher temperatures. Corrosion resistant.	Resists flame impingement. Good resistance at high temperature or in corrosive media.

Table 8 *(concluded)*

Properties	Coating Composition (wt. %) and Designation		
	70% Cr_3C_2+ 30% Nichrome (LC-1D)	80% Cr_2O_3+ 20% Al_2O_3 (LC-5)	99% Al_2O_3 (LA-2; AMS 2436)
Vickers hardness (300 kg. load)	625	925	1100
Maximum temperature in oxidizing atmosphere (F)	1800	1600	1800
Coefficient of thermal expansion (in./in.°F)			$3.8 \cdot 10^6$ (70 ·1830° F)
Modulus of rupture (psi)	95,000	15,000	20,000
Modulus of elasticity (psi)	$21 \cdot 10^6$	$8 \cdot 10^6$	$12 \cdot 10^6$
Porosity (%)	1	1.5	2
Specific Gravity			3.5
Thermal Conductivity (Btu/ft./hr./°F/sq. ft.)			
at 200° F			1.2
at 500° F			0.9
Characteristics of Flame-Plated Coating	Excellent resistance to mechanical and thermal shock at high temperatures.	Good self-mating characteristics; resists wear and chemical attack.	Excellent resistance to wear, chemical attack, and high-temperature deterioration.

coatings of alumina have a low emissivity and insulate well against radiant heat [74]. Alumina coatings also have good resistance to chemical corrosion and mechanical erosion. Optical control coatings of Al_2O_3 have been used on Tiros and Explorer nose cones [14]. Thermally protective coatings of Al_2O_3 have been used for rocket motor components (thrust chamber, nozzle, etc.) in such vehicles as Agena and Bullpup, for jet-engine components (combustion chamber, after burners, turbine blades, etc.) and for numerous furnace components [14, 84, 85]. Al_2O_3 coatings have also been used for protecting and fastening temperature-sensing wires and strain gauges [14, 86]. Because of their high wear resistance, Al_2O_3 coatings have been used for protecting mixing paddles, bearings, grinding media, pouring chutes for transferring molten metals, and as radomes [51, 72, 87, 57].

Zirconia coatings, like coatings of Al_2O_3, have been used primarily for thermal and electrical resistance [82, 88]. The higher melting point of zirconium oxide (4500° F) extends its range of thermal applications beyond that of Al_2O_3. Zirconium-oxide coatings have been used as rocket motor liners and thrust chambers as well as for high-temperature extrusion dies and annealing rolls [77, 51, 14, 54]. The good chemical stability of zirconium oxide also makes it an effective coating for protecting surfaces exposed to molten metals.

Chromium oxide produces dense wear-resistant coatings which are finding increased utilization. These coatings have a low coefficient of friction and good finishing capabilities. Chromium oxide coatings have been used on suction-box covers in paper machines [74], pickup and backup rolls used in the chemical industry, on metal thread grids [3] and on steel rocket-launcher tubes [25].

Flame-sprayed coatings of zircon ($ZrSiO_4$) are hard and adherent and have good resistance to wetting by molten metals. Thermal-barrier coatings can be used to about 3000° F. During spraying, the zircon dissociates into ZrO_2 particles dispersed in a glassy silica matrix and forms coatings which have excellent thermal shock, chemical corrosion and abrasion resistance. Zircon coatings have been used as thermal insulation for rocket sled components [57, 64] and corrosion-resistant crucible linings [14, 76].

In addition to those oxides described, there is a wide range of other oxides which can be applied by flame spraying. Magnesium oxide and $MgSiO_4$ have excellent resistance to molten metals as do the zirconates of barium, calcium and strontium. These materials are often used for coating graphite molds and crucibles [74]. The zirconates and titanates can also be used as dielectric coatings. Beryllium-oxide and uranium-oxide coatings are frequently used for nuclear applications [74, 89]. Cerium oxide and other rare-earth oxides are used as thermal barriers and combustion catalysts in diesel engines [77, 74, 90]. Hafnium oxide and mixtures of hafnium oxide-thorium oxide are used as thermal barriers [91, 93]. One reported application for hafnium oxide-thorium oxide is for coating graphite rocket nozzles [92]. Titanium oxide produces hard coatings with low porosity that adhere well to numerous base materials. Some reported applications of titanium oxide are as wear-resistant coating for golf clubs, and as hard surfacing for aluminum [72, 57].

Flame-sprayed glass and enamel compositions have long been of interest to the ceramic field. The first patents for flame spraying glass coatings date back to the 1920's. In most of the patents the necessity for extented particle heating time or preheating particles and substrate is emphasized.

The high specific heat and low thermal conductivity of most glasses make them difficult to heat and may account for these special requirements. The effect of infrared transparency on the heat transfer process has not been clearly established; however, this may be significant for spraying glass powders. In general the silicates are also difficult to spray because of their high viscosity at the liquidus temperature (~ 1000 poise). The added resistance to shear stresses upon impact limits the deformation and inhibits particle fluidity on the substrate surface. Glasses having high SiO_2 contents are extremely difficult to spray, particularly in the powder form.

The rod spray process is one method employed for spraying compositions having a silica content greater than 20%. Rods are prepared from 10—60% glass (Pyrex or Vycor) with Al_2O_3 and ZrO_2 as the second component. Coatings prepared from these rods were reported to be quite impervious (1—2% porosity) and provided oxidation protection for molybdenum substrates at elevated temperatures for extended periods of time [94].

A number of low-silica glass compositions based on the lead borosilicates and lead-aluminum phosphates have shown considerable promise as coatings for concrete block, steel plates and ceramic tile surfaces. Acid and alkali resistance of these coatings have been enhanced by 5% additions of ZrO_2 and/or TiO_2. In order to achieve good adherence for these coatings to the steel plates it was necessary to heat the substrate to 800° F [95].

Carbide coatings are used primarily for wear resistance and thermal resistance in reducing atmospheres. Cobalt, nickel-chromium and other metals are frequently added in small percentages as fluxes and to improve bonding characteristics. Tungsten carbide is the most commonly used carbide coating because of its excellent resistance to abrasion. Tungsten carbide coatings with 7—15% cobalt are used for wear-resistant seals, cutting edges, hardfacing for dies and bushings and for protection of gas-turbine engine components [74, 57, 40].

Chromium carbide coatings are also used for many wear-resistant applications [98]. Additions of 15—25% nickel-chrome are often used for jet engine components to prevent wear at high temperatures. The higher nickel-chrome compositions tend to have greater resistance to mechanical and thermal shock. Compositions of titanium carbide and tantalum carbide have also been used for wear resistance. Sprayed layers of tantalum carbide have been used as a matrix phase for wire-wound tungsten rocket motors [33]. Zirconium carbide-hafnium carbide compositions have been used for oxidation-resistant coatings because of the coherent glassy film formed by exposure to oxygen at high temperature [91]. Coatings of boron carbide have been used in nuclear control application [14, 77].

Flame sprayed borides are useful in nuclear applications, because boron has a high neutron absorption capacity and good corrosion and erosion resistance. Boride coatings are hard, with good resistance to wear and molten metals [74] Zirconium boride coatings have been successfully used for thermal and oxidation resistance [40, 50].

Silicides yield hard, dense coatings with good wear resistance and good resistance to oxidation due to the self-healing glassy SiO_2 film formed on exposure to air at elevated temperatures. The most commonly used silicide is molybdenum silicide, which can be used to 3100° F in air to protect molybdenum and niobium refractory metals [25, 76, 40, 50]. Aluminum and tantalum silicide have also been used for oxidation-resistant coatings [97, 98].

A wide range of metals and alloys have been flame sprayed, satisfying an extensive list of coating requirements. A partial compilation of the properties and applications of frequently used flame sprayed metals and alloys is present in Table 9.

Metal coatings are often used for building up worn or damaged metal parts, for corrosion protection and wear resistance. Sprayed metal coatings tend to be

Table 9. *Properties and Processing of Flame-Sprayed Metals*
(From MOCK [74])

Type	Aluminum	Babbitt A[1]	Brass (65:35)	Bronze AA[2]
Specific Gravity	2.41	6.67	7.45	7.06
Ult. Ten. Str. (1000 psi)	19.5		12	29
Strain at Ult.-Str. (%)	0.23		0.45	0.46
Rockwell Hardness	H 72	H 58	B 22	B 78
Shrinkage (in./in.)	0.0068		0.009	0.0055
Spraying Speed (lb./hr.)	18	95	32	24
Spraying Efficiency (%)	89	69	81	77
Major characteristics and Uses	Good corrosion and heat resistance.	Good bearing properities.	Sprays fast; fair machine finish.	Hard, very wear resistant; easily machined.

[1] Lead-free, high-tin alloy.
[2] Aluminum-iron-bronze.

Table 9 *(continued)*

Type	Comm. Bronze	Manganese Bronze	Phosphor Bronze	Tobin Bronze	Copper
Specific Gravity	7.57	7.26	7.68	7.46	7.54
Ult. Ten. Str. (1000 psi)	11.5	12	18	13	
Strain at Ult.-Str. (%)	0.42	0.46	0.35	0.51	
Rockwell Hardness	B 18	B 27	B 20	B 27	B 32
Shrinkage (in./in.)	0.011	0.009	0.010	0.0104	
Spraying Speed (lb./hr.)	24	36	31	36	29
Spraying Efficiency (%)	82	79	85	80	80
Major characteristics and Uses	Softest bronze; fair machine finish.	Excellent machine finish; special uses only.	Fair machine finish; special uses only.	General purpose; fair machine finish.	Electrical uses; brazing.

Table 9 *(continued)*

Type	Lead	Molyb-denum	Monel	Nickel	18—8 Stainless	High-Cr Stainless
Specific Gravity	10.21	8.86	7.67	7.55	6.93	6.74
Ult. Ten. Str. (1000 psi)	7.5	21	17.5		30	40
Strain at Ult.-Str. (%)		0.30	0.26	0.30	0.27	0.50
Rockwell Hardness		G 38	B 39	B 49	B 78	C 29
Shrinkage (in./in.)		0.003	0.009	0.008	0.012	0.0018
Spraying Speed (lb./hr.)	80	8	17	18	21	19
Spraying Efficiency (%)	65	87	85	79	81	81
Major characteristics and uses	Good corrosion resistance; X-ray shielding.	Used as bonding coating; excellent bearing properties.	Good corrosion resistance; good machine finish.	Good corrosion resistance; fair machine finish.	High corrosion resistance; good wearing properties.	High hardness and wear resistance; grind finish.

Table 9 *(concluded)*

Type	Steel (LS)	Steel 1010 (0.10% C)	1025 Steel (0.25% C)	1080 Steel (0.80% C)	Tung-sten	Zinc
Specific Gravity	6.78	6.67	6.78	6.36	16.5	6.35
Ult. Ten. Str. (1000 psi)	33.5	30	34.7	27.5	7.0	13
Strain at Ult.-Str. (%)	0.45	0.30	0.46	0.42		1.43
Rockwell Hardness	C 25	B 89	B 90	C 36	A 50	H 46
Shrinkage (in./in.)	0.002	0.008	0.006	0.0014		0.010
Spraying Speed (lb./hr.)	18	19	19	19	5.5	61
Spraying Efficiency (%)	87	87	87	87		66
Major characteristics and uses	Good mechanical and finishing properties.	Simple bearing surfaces and press fits; excellent machine finish.	Harder and lower shrinkage than 0.10 excellent machine finish.	Very hard and wear resistant; good bearing properties; grind finish.	High heat resistance.	Good all around corrosion resistance.

somewhat porous and their oil retention properties provide good bearing surfaces. For other applications the pores can be sealed by a microcrystalline waxy, phenolic or silicone sealer. In general, the metal coatings have lower tensile and torsional strength but higher compression strengths than obtained by other metal fabrication processes. As a rule the density is also somewhat lower.

Aluminum and zinc are commonly used to protect industrial equipment and structures subject to corrosion, such as bridges, lock gates, tanks and marine ware [74, 57, 59]. Protection for fifteen to fifty years of life can be obtained with these coatings. However, care must be exercised to avoid galvanic corrosion (copper in contact with zinc, etc.). Zinc coatings can be used for corrosion-resistant undercoatings for organic materials such as paint and plastic films [74]. For avoidance of galvanic corrosion, tin coatings are often used.

Copper and copper-base alloy coatings are used for a wide range of electrical applications such as carbon brushes and resistors, on glass heating panels for solder connections and on ferrite cores [74, 89]. Nickel and steel coatings are used for rebuilding worn parts, corrosion protection and wear resistance. Other wear or bearing surfaces are prepared from chromium-boron-nickel, nickel-chrome, and chromium-cobalt-tungsten alloys coatings [57, 100]. Nickel-aluminum and silicon coatings are used for oxidation resistance [25, 101]. Nickel-aluminum is also used as an intermediate layer to improve coating adherence.

Numerous applications for refractory metal coatings are also reported in the literature. The refractory metals most frequently used are molybdenum and tungsten. Molybdenum coatings have a high melting point ($4670°$ F), a hard wearing surface and good adhesion to the substrate. These coatings are used to build up worn surfaces of aluminum or iron (brakedrums, grinders, spindles, etc.) as antigalling surfaces and as an interfacial bonding medium [74, 76, 102]. Molybdenum coatings are also used as rocket nozzle liners and to protect jet engine components [19].

Metal and nonmetallic parts exposed to high temperatures are often protected with flame sprayed coatings of tungsten. Thin-walled components of tungsten are also made by the flame spray process. Crucible linings, rocket nozzles and other rocket motor components as well as wear-resistant surfaces are prepared from tungsten coatings [76, 19, 103]. Plasma sprayed coatings of tungsten are also used to protect the graphite components of Polaris missiles [74].

Composite coatings of metal and ceramic mixtures or alternate layers of ceramics and metals have been developed for many thermal, chemical, electrical and mechanical applications. Alternate layers of Al and Al_2O_3 have been used in ultrathermic capacitors as well as for oxidation resistant coatings [97, 104]. $Mo-Al_2O_3$, $NiCr-Al_2O_3$, $NiCr-ZrSiO_4$, $NiAl-Al_2O_3$, $W-ZrO_2$, $NiCr-Al_2O_3$ and Ni-MgO composites have been used for oxidation and thermal resistance in rocket engines, armaments and brake linings [64, 40, 105, 97, 23, 105, 107]. Coatings of molybdenum sulfide with metal binders are used for lubricating surfaces [108]. The numerous composite compositions which can be sprayed offer a large selection of coating systems for a multitude of potential applications.

Composite coatings can be prepared by multigun spraying or with a single gun using multifeed powder injection systems or by using a premixed powder blend in a single-hopper feed system. The simplest and least expensive method is to use premixed powder blends. However, to obtain the desired microstructure and composition on deposition, careful control must be exercised over the spraying process.

The effects of particle size ratio, spray distance, power level, and composition were studied on premixed blends of NiCr and ZrO_2. Coating microstructure, composition, thermal shock resistance and bond strength were evaluated [22].

The effect of the particle size ratios on the coating microstructure of four 15 NiCr-85 ZrO_2 blends is shown in Fig. 44. In the coating (a) prepared from a blend using fine NiCr particles (average diameter 40 μ) and coarse ZrO_2 particles (average diameter 90 μ), very little of the ZrO_2 was melted and most of it appears as irregular particles in a NiCr matrix. The coating prepared from a blend using medium-sized (average diameter 65 μ) NiCr and ZrO_2 particles, (b), is very similar to the first, with only slightly more melted ZrO_2. In the coating (c), prepared from a blend using coarse NiCr particles (average diameter 90 μ) and fine ceramic particles (average diameter 40 μ), the ZrO_2 is much more lamellar and a greater amount is retained. In the coating (d), prepared from coarse NiCr metal particles (average diameter 90 μ) and moderately fine ZrO_2 particles (average diameter 50 μ), the ZrO_2 is retained to a greater degree and the particles appear to be completely deformed; furthermore, the ZrO_2 is now the matrix for the lamellar NiCr phase. These photomicrographs show that the optimum ratio can be exceeded at both ends of the range.

The effect of spray distance upon coating microstructure for an 85 ZrO_2-15 NiCr blend (90 μ NiCr and 50 μ ZrO_2 powders) is shown in Fig. 45. The most significant effect was the increase in ceramic phase retained in the coating with decreasing spray distance. The shorter spray distances produced coatings in which the metal and ceramic particles were more lamellar.

The effect of the input power level in plasma spraying a ceramic-metal composite is shown in Fig. 46. From a comparison of the four photomicrographs, it is apparent that as the power level was raised from 11—21 kilowatts, the amount of ZrO_2 in the coating increased and a more lamellar structure was formed. This is a direct indication of greater melting of the ZrO_2 particles.

The effect of composition was studied on coatings prepared from blends ranging in metal content from 5—100%. Prior to coating, the steel substrates were flashed with a 1—2 mil layer of Mo. Metallographic analysis of the microstructure obtained from coatings prepared from 90 μ NiCr-50 μ ZrO_2 blends of varying metal-ceramic ratio showed that the composition of the feed mixtures was preserved in all of the sprayed coatings. As shown in Fig. 47, both the NiCr and ZrO_2 phases were lamellar and fairly dense. These photomicrographs indicate that controlled microstructures can be achieved by the proper selection of particle size and spray parameters.

The effects of composition on bond strength and thermal shock resistance are shown in Fig. 48. As the percent of ceramic phase increases, the bond strength and thermal shock resistance decreases. Bond strength decreased from 2900 psi to 1900 psi as the ceramic content increased from 65—95%. The thermal shock resistance dropped from 45 cycles to 2 cycles over this same range. It is interesting that as the amount of metal is increased from 5—10% a very marked improvement in thermal shock resistance is observed (from 1 to 19 cycles). Also

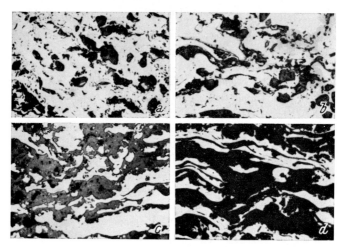

Fig. 44. Effect of Particle-Size Ratio with Metal and Ceramic Powders
(15% NiCr-85% ZrO_2) 250× (From HECHT [22])
(a) 90-μ ZrO_2, 40-μ NiCr
(b) 65-μ ZrO_2, 65-μ NiCr
(c) 40-μ ZrO_2, 90-μ NiCr
(d) 50-μ ZrO_2, 90-μ NiCr

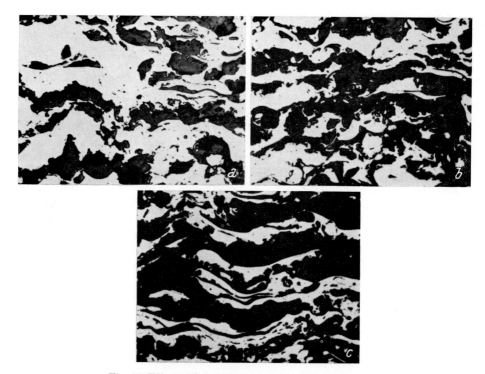

Fig. 45. Effect of Spray Distance (15% NiCr-85% ZrO_2)
(From HECHT [22])

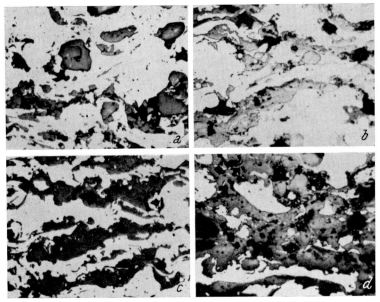

Fig. 46. Effect of Spray Power (15% NiCr-85% ZrO$_2$) 250×
(From HECHT [22])

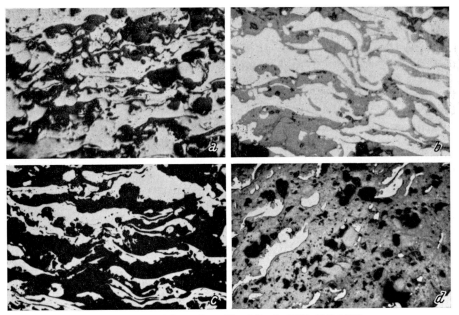

Fig. 47. Effect of Ceramic-to-Metal Ratio (NiCr-ZrO$_2$) 250×
(From HECHT [22])
(a) 35% NiCr-65% ZrO$_2$
(b) 25% NiCr-75% ZrO$_2$
(c) 15% NiCr-85% ZrO$_2$
(d) 5% NiCr-95% ZrO$_2$

of interest was the improved bond strength and thermal shock resistance observed when a small amount (15%) of coarse ceramic particles was dispersed in the metal matrix. This improvement may be due to an annealing effect induced by the heat retention of ceramic particles, which reduces the cooling rate of the sprayed coating. Another possible explanation is that the ceramic particles act as stress-relief sites.

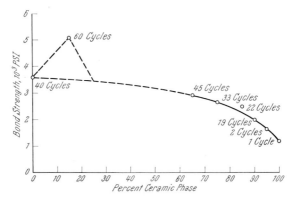

Fig. 48. Bond Strength and Thermal-Shock Resistance Versus Percent Ceramic Phase
(From HECHT [22])

A similar study was conducted with coatings prepared from NiAl-Al$_2$O$_3$ blends. The results reinforced the NiCr-ZrO$_2$ findings, especially regarding the improved bond strength and thermal shock resistance obtained with the addition of coarse particles from the second phase.

The results of this work show that obtaining a desired microstructure for two-phase coatings requires the selection of the proper particle size ratio and the proper spray settings. It is also important to note that there is an interdependency among all three of the spraying parameters studied, i. e., distance, power level, and powder size ratio. In more recent work it has been found that electrode design and the location and angle at which the powders are injected into the effluent are also quite important.

Based on these results, graded composite coatings of NiCr-ZrO$_2$ and NiAl-Al$_2$O$_3$ have been prepared for evaluation. For the NiCr-ZrO$_2$ system, a 2-mil coating of Mo was applied to the substrate followed by three NiCr-ZrO$_2$ graded layers each approximately 3-mil thick, and a 4-mil top coating of ZrO$_2$. For the NiAl-Al$_2$O$_3$ system, a 2-mil coating of NiAl was applied to the substrate followed by three NiAl-Al$_2$O$_3$ graded layers, each 3-mil thick, and a 4 to 5-mil top coat of Al$_2$O$_3$. Bond strength test results for these coatings are summarized in Table 10.

The NiAl-Al$_2$O$_3$ graded coatings had an average bond strength of 2270 psi and failure occurred through the graded Al$_2$O$_3$ zone. The three NiAl-Al$_2$O$_3$ coatings subjected to thermal cycling consistently failed on the 15th cycle. These results correlated closely with the data obtained for the individual blended layers.

The NiCr-ZrO$_2$ graded coatings had an average bond strength of 2560 psi. Coating separation occurred as an irregular shear line which passed through the second and third graded zones. This bond strength level was in close agreement with previously measured values for a blended NiCr-ZrO₂ layer. A photomicrograph of a typical graded NiCr-ZrO₂ coating evaluated for bond strength is shown in Fig. 49.

Table 10. *Results of Bond Strength-Tests for Graded Coatings*

Material System	Sample No.	Bond Strength (psi)	Mode of Failure
NiCr-ZrO₂	1	2580	In graded
	2	2540	zone
	3	2480	
	4	2635	
	5	2565	
	Avg.	2560	
NiAl-Al₂O₃	6	2495	At Al₂O₃
	7	2345	graded-zone
	8	1935	interface
	9	2500	
	10	2075	
	Avg.	2270	

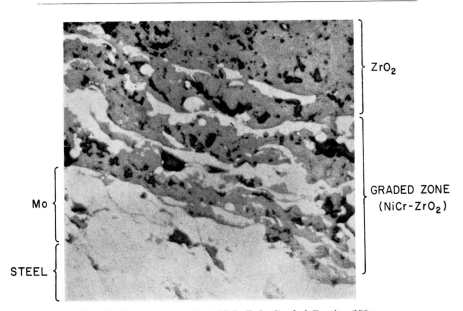

Fig. 49. Photomicrograph of NiCr-ZrO₂ Graded Coating 250×
(From Hecht [22])

The thermal shock resistance of the NiCr-ZrO₂ coatings was disappointing. Failure occurred during the fourth cycle of exposure at a heat flux of 560 Btu/ft² sec. However, these results were better than the results obtained for pure ZrO₂ coatings, which generally failed on the first or second cycle, indicating that some improvement in thermal shock resistance is obtained with graded under-layers.

Analysis of the test specimens showed that failure was due to spalling of the top ZrO_2 layer. One of the test specimens was reexposed to thermal cycling to determine the extent of failure on continued exposures. Although further spalling of the ZrO_2 layer occurred, the graded underlayers did not appear to be affected. This would suggest that the thermal shock resistance of a ZrO_2 layer is significantly improved by small additions of metal particles (15 Ni-85 ZrO_2 withstood 22 thermal shock cycles).

The totally empirical nature of the investigation prompted a study designed to develop a better method for determining the optimum particle size ratio for spray powder mixtures. The study concentrated on the fundamental principles associated with the melting of particles in the plasma effluent. The objective was to eliminate the need for lengthy experimentation each time a new composite system is required. Using the NiCr-ZrO_2 system as a model, the following analysis was developed.

As previously described, ENGELKE [21] developed a model for particle heating in a plasma which gives for a final solution the equation:

$$\left[\frac{S(\lambda \Delta T)^2}{V\eta} \right] = \left[\frac{L^2 D^2}{16\varrho} \right] \tag{4}$$

where:

S spray distance
λ mean boundary-layer thermal conductivity
ΔT mean boundary-layer temperature gradient
V mean plasma velocity
η mean plasma viscosity
L particle heat content per unit volume of the liquid at its melting point relative to the solid at room temperature
D mean particle diameter
ϱ mean particle density

Since the particle temperature is generally a small fraction of the plasma temperature λ, boundary layer conductivity, and ΔT, temperature gradient, across the boundary layer, are governed almost entirely by the enthalpy and composition of the plasma [20]. By rearranging Equation 4, an expression for the spray distance at which the particle will be completely molten, can be written:

$$S = \left[\frac{L^2 D^2}{16\varrho} \right] \left[\frac{V\eta}{(\lambda \Delta T)^2} \right] \tag{5}$$

If a composite coating is to be obtained, the particles for each constituent phase will have to be molten within the fixed spray distance. Ideally, all of the particles injected into the effluent should become completely molten just prior to substrate impact. Thus, for the cermet coating system NiCr-ZrO_2, the spray distance for the NiCr particles (S_N) should be the same as for the ZrO_2 particles (S_Z).

$$S_N = S_Z \tag{6}$$

Therefore:

$$\left[\frac{L^2 D^2}{16\varrho}\right]_N \left[\frac{V\eta}{(\lambda \Delta T)^2}\right] = \left[\frac{L^2 D^2}{16\varrho}\right]_Z \left[\frac{V\eta}{(\lambda \Delta T)^2}\right] \tag{7}$$

Since the effluent conditions are identical for both particles, Equation 7 simplifies to:

$$\left[\frac{(LD)^2}{16\varrho}\right]_N = \left[\frac{(LD)^2}{16\varrho}\right]_Z \tag{8}$$

The particle heat content per unit volume can be calculated from the relationship:

$$L = \varrho\,[C_p \Delta T_m + H_f] \tag{9}$$

where:

ϱ particle density
C_p specific heat
ΔT_m increment of temperature required to melt the particle
H_f heat of fusion

The equation for L can be substituted into Equation 8 to give:

$$\frac{[\varrho(C_p \Delta T_m + H_f)]_N^2\, D_N^2}{16\varrho_N} = \frac{[\varrho(C_p \Delta T_m + H_f)]_Z^2\, D_Z^2}{16\varrho_Z} \tag{10}$$

This can be simplified to:

$$D_N^2 \varrho_N (C_p \Delta T_m + H_f)_N^2 = D_Z^2 \varrho_Z (C_p \Delta T_m + H_f)_Z^2 \tag{11}$$

Equation 11 can be further rearranged to give particle size ratio as a function of material density and energy required for particle melting.

$$\frac{D_N}{D_Z} = \frac{\sqrt{\varrho_Z}\,(C_p \Delta T_m + H_f)_Z}{\sqrt{\varrho_N}\,(C_p \Delta T_m + H_f)_N} \tag{12}$$

Substituting the appropriate property values for NiCr and ZrO$_2$ in this relationship gives the particle size ratio:

$$\frac{D_Z}{D_N} = 1.72 \tag{13}$$

Based on these calculations the optimum diameter for the NiCr particle would be 1.72 times larger than the diameter of the ZrO$_2$ particle. Thus, if a diameter of 50 μ is selected for the ZrO$_2$ particle, a diameter of 86 μ would be required for NiCr particle. The empirical analysis for the NiCr-ZrO$_2$ blend showed that for a 50 μ diameter ZrO$_2$ particle, the optimum diameter for a NiCr particle was 90 μ. The close agreement between the calculated values and the experimental values was encouraging.

Four metal-ceramic systems were selected for further evaluation of the above equations, developed for determining the optimum particle size ratios. The thermal and physical property data for each material are listed in Table 11. The

Table 11. *Property Data for Components of Some Metal-Ceramic Systems*

Material	ϱ (lb./ft.^{3})	ΔT_m (°F)	C_p $\left(\dfrac{Btu}{lb.\ °F}\right)$	H_f (Btu/lb.)
Al$_2$O$_3$	247.8	3700	0.21	508.2
ZrO$_2$	329.6	4800	0.15	303.4
SrTiO$_3$	319.0	3700	0.12	156.7
SiZrO$_4$	293.4	3400	0.20	215.0
Ni	555.6	2650	0.11	129.2
NiCr	536.9	2600	0.13	130.1
NiAl	368.3	2980	0.12	168.0

Fig. 50. Photomicrographs of Theoretically Formulated Metal-Ceramic Blends 250×
(From HECHT [22])
(a) NiAl-Al$_2$O$_3$
(b) NiCr-ZrO$_2$
(c) Ni-SrTiO$_3$
(d) Ni-SiZrO$_4$

particle size ratios for the four composite systems calculated from these data are presented in Table 12.

Using these values, powder spray blends of 50% ceramic phase and 50% metallic phase (by volume) were prepared. Metallographic samples of each system were plasma-sprayed on steel specimens for evaluation. The photomicro-

Table 12. *Optimum Particle Size Ratios*

Systems	
1. NiAl-Al$_2$O$_3$	D(Al$_2$O$_3$) = 0.6 D(NiAl)
2. NiCr-ZrO$_2$	D(ZrO$_2$) = 0.7 D(NiCr)
3. Ni-SrTiO$_3$	D(SrTiO$_3$) = 1.2 D(Ni)
4. Ni-SiZrO$_4$	D(SiZrO$_4$) = 0.9 D(Ni)

graphs in Fig. 50 show that the coatings are generally lamellar and approximate the presprayed composition. The striated microstructure observed for the Ni-SrTiO$_3$ coating system (Fig. 50C) is obtained by special traversing techniques and is included to demonstrate the degree of microstructure control which can be achieved.

These studies showed that plasma sprayed composite coatings can be tailored to meet specific requirements. Predetermined volumetric proportions and microstructures in a binary coating can be achieved and coating properties, such as bond strength and thermal shock resistance, can be modified. Fabricating the desired composite microstructure requires control of the critical process variables: powder particle size ratio, spray power, and spray distance. It is also important to recognize the interdependency among these process variables. Heat transfer from the plasma effluent to the particle appears to have the most critical relationship to the thermal and physical properties of the materials and it is directly affected by the selection of the particle size ratio.

5. Material Evaluations Utilizing the Plasma Jet

It is only natural that a device such as the arc plasma jet, which is capable of generating on a continuous basis an effluent which has far higher temperatures and produces higher heat fluxes than any other known technique, should get widespread use in materials testing. Test facilities utilizing the unique capabilities of the plasma jet are in extended use for many different and widely varying types of material evaluations. Among the more common of these are ablation, reentry simulation, thermal shock, thermal stress, dynamic oxidation and rocket exhaust simulation. The techniques and degree of sophistication of these evaluations range from the simplest of screening tests to highly detailed simulations of conditions which would exist in an actual application.

5.1. Plasma Test Facilities

A limited survey [110] of typical test facilities in which plasma jets are used as heat sources is presented in Table 13. This sampling demonstrates the extensive variety in the types of basic equipment, capabilities and techniques which are employed. Five different methods of arc stabilization are listed with several units making use of more than one method. Electrode materials include copper, tungsten, carbon, graphite, and silver-copper alloys. A tungsten rear electrode (cathode) in conjunction with a water-cooled copper front electrode (anode) is widely used with all stabilization techniques in the low to medium power ranges. The units capable of high power operation often require the use of water-cooled copper for the cathode as well. This reduces the contamination of the test effluent which occurs if the tungsten cathode is overheated and ablates. Carbon and graphite cathodes likewise produce more contaminated streams and therefore are less popular.

The listed facilities range in power levels from 40 kilowatts to 25 megawatts. Nozzle exits are up to 48 inches in diameter and velocities range from subsonic to Mach 18. Although the majority of units are powered on direct current, five of the facilities surveyed used 60 Hz alternating current and one has the plasma generated by radio-frequency inductive coupling. Rectifiers are the most widely used method for producing the direct current power. The alternating current is obtained from transformers and the radio frequency from high frequency transmitters.

Although there are general problem areas common to all facets of plasma testing, each individual type of test and test facility has its own unique characteristics and problems. It is therefore well to review and describe in some detail several typical types of evaluations which are conducted in arc plasma facilities.

Table 13. *Typical Arc-Plasma Test Facilities and Capabilities*
(Ref. 110)

Facility	Power Supply	Electrodes	Method of Arc Stabilization	Exit-Nozzle Diameter	Mach Number
Johns Hopkins University	12.5-MW Submarine Battery	Copper/ Copper	Magnetic	0.25—16 inches	1—10
Arnold En- gineering Development Center	40-KW Rectifier		Constrictor	2—6 inches	4—14
Avco Corp.					
#1	500-KW Rectifier	Tungsten or Carbon/Copper	Vortex	0.5 and 0.6 inch	0.3—0.8
#2	250-KW Rectifier	Carbon/ Copper	Vortex	0.5 and 0.6 inch	0.5—0.6
#3	2-MW Rectifier	Tungsten/ Copper	Vortex	3—5 inches	3—5
#4	10-MW Batteries	Carbon/ Copper	Vortex	0.4—12 inches	0.4—5
Boeing Company	160-KW Rectifiers			2 inches	Subsonic
	1-MW Generator			1—8 inches	2.4—11
Fairchild Hiller	1-MW Rectifier	Tungsten/ Copper	Vortex	0.78—2.5 inches	2—5
General Dynamic	1.8-MW Rectifier	Copper/ Copper	Magnetic	1.34—2.68 inches	3
General Electric					
#1	25-MW Transformer (AC)	Copper/ Copper		0.5 inch	1—8
#2	500-KW Rectifier		Vortex or Wall	0.5—1.2 inches	Up to 8
Giannini Scientific	1-MW Rectifier		Vortex	0.836—5.25 inches	2.5—3.5
Ling-Temco- Vought	180-KW Rectifier	Tungsten/ Copper	Vortex	0.84—1.5 inches	2.5—3.0
Martin- Marietta					
#1	240-KW Rectifier	Tungsten/ Copper	Vortex	1.5—3 inches	3
#2	1-MW Rectifier	Tungsten/ Copper	Vortex	0.5—15 inches	1.5—7.0

Table 13 *(continued)*

Facility	Power Supply	Electrodes	Method of Arc Stabilization	Exit-Nozzle Diameter	Mach Number
Marquardt	6-MW Batteries	Copper/ Copper	Magnetic	2—8 inches	3—11
McDonnell Aircraft	3.4-MW Rectifiers	Copper or Copper Silver Alloy/Copper	Vortex and Magnetic	1.5—8.0 inches	Up to 5
NASA, Ames					
#1	2.5-MW Motor Generator	Tungsten/ Copper	Wall	6.3 inches	4
#2	15-MW Rectifier	Tungsten/ Copper	Wall	5.9—17.7 inches	4
#3	2.5-MW Generator	Tungsten/ Copper	Wall	2.38 inches	3—5
NASA, Langley					
#1	20-MW Rectifiers	Copper/ Copper	Magnetic		12—18
#2	1.5-MW Motor Generator	Silver Copper Alloy/Copper	Magnetic		12
#3	15-MW 3-Phase A.C.	Copper/ Copper		1.75—6 inches	Up to 1.8
#4	15-MW Rectifiers	Copper/ Copper	Gas Sheath	1—4 inches	2—6
#5	15-MW 3-Phase A.C.	Copper/ Copper		2—6 inches	2—4
NASA, MSC					
#1	1-MW Rectifiers	Tungsten/ Copper	Vortex and Magnetic	3.0 inches	Subsonic
#2	1.5-MW Rectifiers	Tungsten or Copper/Copper	Gas Sheath and Magnetic or Vortex	3—8 inches	0.2—5.0
#3	10-MW Rectifiers	Copper/ Copper	Vortex and Magnetic		0.2—10
North American Aviation	1-MW Rectifiers	Tungsten/ Copper	Vortex	0.25—3.5 inches	Up to 3
U.S. Army Redstone Arsenal	18-MW Generators	Copper/ Copper	Magnetic	5—26 inches	Up to 5
U.S. Air Force Systems Command	4.7-MW Rectifiers	Silver-Copper Alloy/Copper	Vortex and Magnetic	7.3—2.4 inches	6—16.5

Table 13 *(continued)*

Facility	Power Supply	Electrodes	Method of Arc Stabilization	Exit-Nozzle Diameter	Mach Number
Stanford Research Institute					
#1	160-KW Rectifiers	Tungsten/ Copper	Vortex	0.69 inch	Subsonic
#2	50-KW RF Transmitters (4—13 mc)	N.A.*			Subsonic
U.S. Naval Ordnance oratory	3-MW 3-Phase A.C.	Copper/ Copper		1.8 inches 6 inches 10 inches 14 inches	4.5 7 9 11
University of Dayton Research Institute					
#1	120-KW Rectifiers	Tungsten/ Copper	Vortex	0.5 inch	Subsonic
#2	50-KW Rectifiers	Tungsten/ Copper	Vortex	0.10—0.15 inch throat**	Sonic throat
University of Minnesota***	250-KW Rectifiers 500-KW Batteries	Copper, Graphite, Tungsten/ Copper	Vortex, Wall, Constrictor		
Westinghouse	Up to 20-MW AC and DC	Copper/ Copper		0.125—48 inches	Up to 16

* Induction.
** Tests nozzle-shaped specimens.
*** Describes several arc heater units.

5.2. Screening of Ablation Materials

Ablation cooling has proven to be an extremely effective technique for temporary thermal protection of structural elements in many aerospace applications. The term ablation has been defined by ASTM as "a self-regulating heat and mass transfer process in which incident energy is expended by sacrificial loss of material" [111]. The sacrificial loss of material is the primary mode by which energy is dissipated in the ablation process. Energy in the form of heat can be absorbed, blocked and/or dissipated by: 1) phase transitions (melting, vaporization and sublimation), 2) mass transfer into the boundary layer, 3) convection in a liquid layer, 4) conduction into a solid body, 5) reradiation, 6) chemical reactions and, 7) mechanical erosion. Each of these mechanisms is reasonably well understood and can be handled analytically without great difficulty [112]. Most materials however, take advantage of more than one of these mechanisms of ablation to accomplish their mission. Furthermore, the mechanisms of ablation which a material exhibits depends not only on the materials composition but also on the specific environmental conditions.

Ablative materials have been employed extensively for thermal protection of nose cones on ballistic reentry vehicles and in various components of rocket propulsion systems. They are also useful in less severe environments such as on lifting reentry vehicles. Here they are used as a coating to protect an underlying load-bearing component during the short period of peak heating which occurs upon initial atmospheric entry.

While reinforced plastics are most commonly associated with ablation systems, a variety of plastic, ceramic, metal, and composite materials have also been employed for this purpose. In the area of reinforced plastics alone, the number of composites which could be fabricated into potentially effective ablative materials is enormous. Considering also all the possible variations in processing and in different forms of reinforcement (fiber, flake, powder, or cloth), it becomes apparent that simple screening tests to evaluate the potential of experimental ablative materials are necessary in the early stages of a materials development program.

As a rule, the environment in which an ablative material must ultimately perform is too complex to be simulated in simple laboratory tests. It is often necessary to compromise by partially simulating end-use conditions and then ranking materials by their performance relative to arbitrary standards. Generally this compromise is accomplished by omission of those parameters which are difficult to simulate, such as high mass flow and high velocity pressure. The screening test often reduces to a duplication of anticipated thermal conditions with omission of the shear forces which also can act upon the surface of an ablating body and strip away material before it reaches full effectiveness in blocking and dissipating input energy. If a screening test does not realistically simulate the proposed application, there is a risk of calling a bad material good. The conservative approach to analysis of screening test results lies in rejection of obviously inferior systems and retention of those materials which exhibit satisfactory behavior for more sophisticated testing [113].

Screening tests for ablative materials represented the first widespread application of the arc plasma jet for the evaluation of high temperature materials. At the time when these initial efforts were being undertaken, nearly all of the plasma jets commercially available were in the 40—50-kilowatt power range. As the demands for larger test sections and more realistic simulations increased, more powerful units were developed and adapted to this use. As a general rule, however, test facilities which are used exclusively for screening utilize plasma jets in the 40—250-kilowatt range.

The plasma jet is usually stabilized with nitrogen (to prevent oxidation of the anode) with provisions for injection and mixing of oxygen in suitable proportions to simulate an air effluent. The oxygen is either injected through the arc column or added in a mixing chamber located downstream from the cathode or front electrode. Some plasma jets employ materials which do not oxidize excessively for both the anode and cathode, and can therefore be stabilized on air. In the later generation screening facilities a vacuum test chamber is sometimes attached to the exit nozzle of the plasma jet and the effluent is expanded into this low pressure tank. The addition of the low pressure capability makes it possible to simulate high altitude conditions, and to obtain hypersonic gas velocities and higher gas enthalpies [114].

The general procedures followed in conducting screening tests do not vary appreciably from one facility to another. The plasma jet is "ignited" and the power level and mass flow rates adjusted to produce the desired test conditions. The test specimen is then inserted into the effluent and exposed for either a predetermined length of time or until some criterion of failure (or success) is met. The specimen is then removed from the effluent and permitted to cool. The evaluation of a particular specimen's performance is usually based on physical measurements which are made before and after exposure, data recorded during the test, and visual observation of the general specimen behavior with emphasis on any undesirable performance characteristics.

Fig. 51. Arc Plasma Splash-Type Screening-Test Facility

Test specimens include flat panels in varying sizes, cylindrical rods, thin discs, cones, wedges, hemispheres and pipes. Fig. 51 shows a flat panel mounted at a 45° angle in a simple "splash-type" screening test. Although it is usual to choose a geometric shape which will result in a desired surface heating rate or the proper flow conditions, simplicity, adaptability and ease of specimen preparation are also taken into account. The instrumentation used to monitor test conditions and specimen response is also widely varied from facility to facility. For that matter, even the parameters which are measured differ. The minimal parameters which are measured in the majority of screening tests include electrical power, energy losses to the coolant, gas flow rates, heat flux and stagnation pressure as well as front and back surface temperature of the specimen. Virtually all screening tests define and establish the test conditions by a measure of heat flux rather than effluent enthalpy or temperature.

5.3. Thermal-Shock Evaluations

There are numerous high-temperature materials applications in which the temporary protection and continually changing physical dimensions of an ablator cannot be tolerated. Refractory ceramics are the most readily available and attractive class of materials for these applications. This category of ceramic materials includes the oxides, carbides, borides, nitrides, aluminides, silicides and sulfides [115]. They are generally characterized by high melting temperatures, relatively low thermal and electrical conductivities, and good stability at elevated temperatures. Ceramics have been used in bulk form as both structural and non-structural (liner) components and also as coatings which are applied to load-bearing members. The oxides are especially attractive since they are not only strong at high temperatures but they are chemically stable in an oxidizing atmosphere.

The susceptibility of ceramics to mechanical failure due to thermal stresses and thermal shock has been one of the principal factors limiting their use. There are many applications for which ceramics possess the required high-temperature structural properties but in which failures would occur at much lower temperatures, such as during heating or cooling [116].

Thermal shocking of a material is brought about by a large, rapid change in the temperature of the surface of a solid body. In less insulative materials such as ordinary metals, these rapid changes in temperature cause no appreciable temperature gradients and therefore no large thermal stresses are generated. This is, of course, due to the high thermal conductivity of metals and their ability to distribute the incident energy rapidly. Refractory ceramics, on the other hand, are usually poor thermal conductors and very large temperature gradients result from a change in surface temperatures.

Even though the materials properties and other factors, (such as shape, density, etc.) are well understood, it is difficult to predict quantitative thermal shock data from purely analytical considerations. Generally, only gross differences in resistance to thermal shock, such as exist between completely different classes of materials, can be predicted. Also, it is often difficult to establish an analytical model of the thermal conditions which a hardware item is likely to encounter. For these reasons the problem of thermal shock can usually only be solved by a careful combination of theoretical principles and experimental evaluation. The result is therefore an empirical solution.

Many types of thermal shock tests have been devised and utilized by the different industries over the years. With very few exceptions these tests consisted of heating a specimen to a desired temperature and then quenching it rapidly by immersing it in a liquid bath. Specimen shapes have included spheres, cylinders, plates, cubes, bars and wedges [117]. These tests have all shared one shortcoming: they are unable to determine a numerical value of thermal shock resistance which could be generally useful. In each case, the test was capable only of generating an order of merit, i. e., the order in which the different materials fall in that particular test. The data generated are therefore only applicable to similar configurations and conditions of heat transfer.

The thermal shock tests conducted in arc plasma jet facilities fall into this same category. However, for potential aerospace applications, they do offer several important differences from most of the other simple thermal shock tests. First, the plasma tests produce the shock conditions by rapid heating of the specimen surface rather than by rapid cooling. Second, the heat transfer conditions (temperature, enthalpy, and the heat transfer coefficient) are similar to those anticipated in the actual application. Another major advantage is the lack of dependence on a particular geometric shape to produce useful information.

Fig. 52. Tube Cluster from Rocket Engine Prepared for Thermal-Shock Evaluation

In conducting a thermal shock test, the plasma effluent conditions are usually defined either by establishing certain gas enthalpies or temperatures or by adjusting conditions to yield a predetermined heating rate. Once the conditions have been established, the specimen is inserted as rapidly as possible. The specimen is often withdrawn, permitted to cool and reinserted many times to determine the effects of cumulative damage. Some tests have been devised wherein the test conditions are made progressively more severe with increasing numbers of cycles. Test specimens are usually similar in configuration to some end-use item. Particular attention is given to simulating or duplicating conditions which may affect the thermal-shock performance of a material, e.g., internal cooling. An actual section from the regeneratively cooled nozzle of the X-15 rocket engine is shown undergoing thermal shock tests in Fig. 52.

The index of performance of such thermal shock evaluations is drawn from observation of physical integrity during exposure. In most cases it is also desirable to record pertinent operational data such as the surface temperature during heating and cooling and also the parameters which describe the properties of the effluent. These data are often invaluable in analyzing a material's performance.

5.4. Thermal-Stress Tests

The state of stress within a material which causes failure in thermal shock also can result in what is more properly termed thermal stress failure. This stress state is, of course, the result of a temperature gradient within the material. The difference between thermal shock and thermal stress is that in the former case the temperature gradient exists primarily because of a rapid change in the tem-

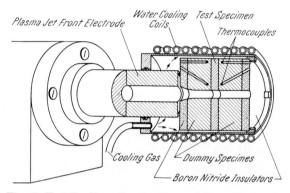

Fig. 53. Test Configuration for Thermal-Stress Evaluation

perature of a portion of the material (such as at the surface) whereas in the latter case the temperature gradient exists primarily because of the resistance to flux of heat from one portion of the material to another. In other words, thermal shock results from a time-dependent temperature differential, whereas thermal stress is independent of time.

While the differences in these two may seem slight, the methods, for evaluating the resistance of materials to failure by these mechanisms are not at all similar. In the thermal shock test using the arc plasma, a large pulse of heat is impacted on the surface, resulting in a rapid rise in temperature. In thermal stress tests, the heat is applied much less harshly, the objective being to establish and sustain a very high rate of heat transfer through the material. The difficulties are therefore compounded, since energy must not only be added at a steady, high rate but it must also be removed from another surface at an equally steady, high rate.

Thermal stress tests conducted in an arc plasma facility utilize techniques which are very similar to those used in nearly all such tests. The specimen is usually in the form of a thick-walled cylinder which is heated on the inside and cooled on the outside. The radial heat flow in this type of specimen results in the required large temperature gradient. In a one-dimensional case of heat transfer, such as heat flowing through a semi-infinite flat plate, the temperature varies directly with the thickness. In the wall of a cylinder the temperature varies as the square of the radius. Since all stresses which develop are hoop stresses, no additional constraints are required. The length of the cylindrical specimen is kept short compared to the other dimensions. Dummy specimens of the same material as that being evaluated are placed on either end of the test specimen to minimize axial heat transfer, uneven heating and nonuniform stresses. Any instrumentation which may be used to monitor specimen conditions is usually placed in these dummy specimens. In this way the actual specimen remains free of any disturbances which could effect its performance.

The dummy and test specimens are attached directly to the plasma jet with the plasma effluent passing through the inside of the cylinder (Fig. 53). A cooling gas such as air, nitrogen or carbon dioxide is passed around the entire outer surface of the cylinder. The enthalpy of the plasma gas is increased gradually until specimen failure occurs. Temperatures measured at different radial distances within the wall of the cylinder are used to calculate the maximum temperature differential which the material could sustain without fracture.

5.5. Dynamic Oxidation

A number of metals and alloys have sufficiently high melting points and also maintain good mechanical properties at temperatures above $2000°$ F. These metallic materials are used in a variety of applications which could not be satisfied by other high temperatures materials such as ablators or bulk refractory ceramics. Sheet panels located on non-stagnation regions of ballistic missiles, the entire skin structure of glide reentry vehicles, inlet diffusers of high-speed air-breathing engines, rocket-nozzle skirt extensions and turbine blades are prime examples of these applications. The primary materials used are the nickel- and cobalt-based superalloys and the refractory metals, niobium, tungsten, tantalum, molybdenum and hafnium and their alloys. Generally speaking, components designed and constructed of these metals rely on reradiation of incident heat for cooling, although other means, such as internal cooling of turbine blades, are used when practical.

All of the super alloys and refractory metals are susceptable to damage by oxidation. Although the degree of susceptibility varies drastically from one alloy to another, it always increases with increasing temperatures [118]. In spite of the problem of oxidation, considerable effort has been expended to render these materials useful at higher and higher temperatures. Coatings of oxidation-resistant materials have been applied to the exposed surfaces of these metals to protect them. Over the past few years inorganic coating technology has advanced tremendously and most coatings are now very often complex combinations of metallic and nonmetallic materials. Different coatings have been developed for the alloys depending on substrate composition, intended application, applicability of processing equipment and economic considerations. The number of coating-alloy combinations which has evolved is very large.

There are other refractory materials which are also subject to damage by oxidation. These include borides, nitrides, carbides, graphite and carbon, as well as a large number of the refractory composites. As these bulk materials and coatings are undergoing development they are subjected to various tests to evaluate their oxidation resistance. The most elementary of these tests and the one which is most widely used is the cyclic oxidation or air furnace test. While this test is very simple to conduct, it has the obvious shortcoming of not simulating the dynamic or moving fluid conditions which exist in nearly all applications. It is also generally limited to temperatures below approximately 3200° F. The high temperature capabilities of the plasma jet, coupled with the ability to control effluent composition, have made it a useful tool in conducting these basic dynamic oxidation studies.

The techniques used in conducting dynamic oxidation tests are similar to those used in simple ablative screening tests. The specimen is inserted in the effluent (usually air) where it remains either for a predetermined length of time or until the coating fails. Cycling techniques are sometimes used where the specimen is removed from the effluent and permitted to cool to room temperature after a prescribed period of exposure, usually one hour. Specimens are weighed before being subjected to the succeeding heating cycle. Test conditions are defined by either a heat flux level or by a desired specimen surface temperature. The surface temperature is usually preferred since the heat flux is representative of an actual application only if the conditions which regulate the reradiation of heat are also simulated. These conditions (shapes, relative positions, absolute temperatures, etc.) are frequently very difficult to achieve.

Metal specimens with coatings are usually in the form of small rectangular tabs or coupons $\frac{1}{2}$ to 1 inch wide by $1\frac{1}{4}$ to 2 inches long. One-inch diameter disc-shaped specimens have also been used on occasion. Specimen thickness varies between 0.020 inches and 0.100 inches. Solid refractory specimens are evaluated in a variety of shapes including rods (mounted normal to direction of flow), cylinders, flat plates and rectangular bars. Both types of materials, coated metal and bulk refractory, are mounted in water-cooled holders whenever possible. Occasionally this form of mounting is impractical because the large amount of heat which is transmitted to the holder makes it difficult to achieve the desired specimen temperature and also introduces large temperature gradients across the specimen. In these cases a refractory material such as an oxide ceramic is used to mount the test sample.

Dynamic oxidation tests are considered simple screening tests designed to generate preliminary data and to eliminate the obviously poor materials from further, more advanced (and much more costly) evaluations. The basis for analysis of a material's performance is therefore also kept simple. For coated metals the criterion of performance is either:

1)   The time at temperature which results in a coating failure or,
2)   The change in specimen weight as a function of exposure time.

Bulk refractories are evaluated on the basis of:

1)   Surface recession rate,
2)   Depth to which material has been affected by oxidation or,
3)   Weight change.

5.6. Reentry Simulation

The test facilities described in the preceeding sections are intended for use as a research tool to assist primarily in the development of improved high temperature materials. Ultimately, however, it is necessary to evaluate both materials and entire system concepts in environmental conditions which more closely resemble those of an actual application. The environment of the intended application, of course, dictates the type of simulation facility which will be most useful. Just as there are varying degrees of sophistication in screening tests, there are varying degrees of simulation in these advanced tests. With additional and more precise simulation the cost of testing increases drastically. A simple ablative screening test can be conducted for $3.00 to $10.00 per specimen, whereas an ablative test in a facility which closely duplicates the conditions encountered during reentry can easily cost 100 times that amount [110, 114]. Less stringent requirements on simulation results in a cost somewhere between these two extremes.

The most elementary and least expensive reentry simulations are accomplished by using a screening facility which has been altered or modified in such a way as to produce the desired conditions. An example of this is the modification to the basic dynamic oxidation test equipment and procedures to partially simulate the conditions encountered by a coated metal panel on a non-stagnation region of a lifting vehicle during a boost-glide-reentry mission [119]. The two primary conditions which are required for this simulation are the varying thermal input and pressure. The requirements for a thermal profile can be easily achieved by simply varying the enthalpy of the effluent by adjusting the power level. Since most small plasma jets, such as those which are used in screening facilities, have large operating envelopes, many time-temperature profiles can be produced in this manner.

Duplication of the reduced-pressure atmosphere encountered at the higher suborbital altitudes could not be accomplished with a common dynamic oxidation facility since these tests are conducted at atmospheric pressure. However, since the materials of interest (oxidation-resistant coatings) fail by reaction with the oxygen in the atmosphere, a partial simulation can be achieved by reducing the amount of oxygen present in the test environment. This is accomplished by reducing the amount of oxygen added to the nitrogen arc gas so that the partial pressure of the oxygen is identical to the oxygen partial pressure at the desired altitude.

However, studies of room-air entrainment [120] on a small plasma unit (50 KW) indicate that the oxygen content of the effluent is quite high, even when the plasma is operated on pure nitrogen. Typical O_2 content in a nitrogen-stabilized plasma effluent emitted from a $1/2$-inch diameter nozzle will vary from 2% at the center of the stream to 7% at a radial distance of $1/4$-inch from the center line. These values were recorded at an axial distance of $1^1/_8$-inches from the exit plane of the torch.

To simulate an altitude of 100,000 ft. requires an oxygen partial pressure of 0.0314 psi, which is equivalent to 0.22% O_2 at one atmosphere pressure. It is evident, therefore, that the amount of air entrained into the plasma effluent must be reduced. To accomplish this, a water-cooled shroud (Fig. 54) consisting of a 2-inch diameter tube approximately 4 inches long was designed to fit around the exit of the plasma head and extend to within $1/4$ inch of the test specimen. This device was fitted with injection ports on the upstream end for the addition of nitrogen cover gas to block the entrainment of room air.

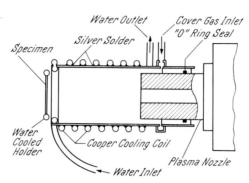

Fig. 54. Shroud to Control Atmosphere for Dynamic Oxidation Tests

With this shroud installed and the plasma jet operating on nitrogen, with sufficient injection of nitrogen cover gas, the oxygen content of the free plasma discharge could be maintained at 0.2% along the center of the stream and 0.3% at a radial distance of $1/2$ inches from the center line. These values could be maintained over the entire range of power levels anticipated for this type of test. The addition of the shroud also caused the plasma "flame" to become more widely dispersed. This reduced the sharp temperature gradient normally present across the plasma effluent and produced a larger cross-sectional area of nearly constant temperature.

Obviously this is not a realistic simulation because the increased presence of the nitrogen in this effluent decreases the diffusion of oxygen at the surface of the test specimen. However, this type of test can yield more information than the simple dynamic oxidation test and the cost is only slightly over that of the simpler evaluations.

Another example of upgrading a simple screening test to achieve better simulation is the use of *spin tests* [121]. A very serious deficiency, particularly with a subsonic screening test, is the lack of appreciable shear stresses on the surface. The high stagnation pressure and velocity encountered on a reentry vehicle results in very large shear forces on the protective surface. These large shear forces drastically alter the performance of materials such as ablators which rely on the formation of a char layer or a viscous liquid phase on the surface as a protective mechanism.

In the spin test, the high aerodynamic shear stresses are simulated by rotating a cylindrical specimen about its axis while exposing the flat end of the cylinder to the plasma stream. The centrifugal forces of rotation acting on the ablating surface result in shearing forces which can be easily regulated. With this test configuration the shear forces cannot always be controlled but they vary predictably across the surface being evaluated.

The spin test has its limits as a simulation of aerodynamically induced shear forces. It neglects the effect which pressure itself has on the ablative performance of materials. However, as in the previous example cited, the spin test capability can be added to a standard screening test at a fraction of the cost which would be involved in conducting tests in a high-pressure, high-shear wind tunnel. It should be noted also that the spinning technique has been used in conjunction with more advanced supersonic test facilities to produce the very high shear forces which would be encountered by a body with extremely high velocity and a steep angle of reentry [121].

For more closely simulated conditions of reentry, an arc-heated or *hyperthermal wind tunnel* is required. These facilities consist basically of an arc plasma jet which exhausts into a reduced pressure chamber (Fig. 55). The source of high-

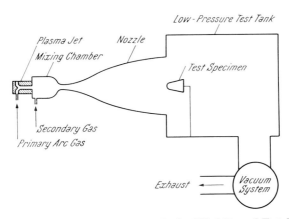

Fig. 55. Schematic of Typical Hypervelocity Wind-Tunnel Test Facility

pressure, high-enthalpy, gas which is required in the wind tunnel is the arc plasma jet. The arc-heated primary gas is passed from the plasma jet into a plenum chamber where cooler secondary gases are added and mixed. This chamber also serves as a stilling section in which the large property gradients which exist in the effluent as it is emitted from the plasma are eliminated, thus forming a uniform, high-energy fluid. From this section the working gas is passed through a supersonic converging-diverging nozzle where it expands with the accompanying increase in velocity. The test model is usually placed at or near the exit of this nozzle section. A vacuum test chamber is fitted to the exit of the nozzle to collect the expanded gas. Large vacuum pumps or some other means of maintaining the low pressure in this tank (such as steam ejectors) are required.

With a properly designed nozzle and sufficiently low tank pressure it is theoretically possible to expand the gas to a condition of ambient static temperature and velocities up to many times the speed of sound. This should result in conditions at the test site which would be very similar to those encountered during an actual hypersonic flight. Generally speaking however, this is not practical. Most test facilities simulate the stagnation enthalpy and stagnation pressure rather than the free-stream Mach number [1]. This can be justified since the parameter of major concern, heat transfer to the specimen, depends primarily on duplicating the stagnation or total conditions of enthalpy and pressure.

Although the hyperthermal wind tunnels are by far the best available means of simulating reentry conditions they are also not without limitations. For example the majority of facilities currently in operation are not capable of programming variations in test conditions as a function of time to simulate an actual reentry in its entirety. They are generally limited to simulating only a portion of the reentry profile or even one specific set of conditions during any one test. Another difficulty arises in the equilibrium condition of the air as it reaches the test site. In a plasma tunnel the chemical composition of the air at this point does not match the equilibrium values associated with local temperatures. This nonequilibrium condition usually maintains itself over the entire test section. Chemical reactivity at the surface of a test model will therefore not duplicate that which occurs on an actual flight vehicle [1]. One of the many problems of chemical reactivity is the ion-electron and atom recombination at the surface which results in considerable energy release. This particular problem will be discussed in greater detail later.

The parameters such as enthalpy and pressure, which are usually duplicated in a reentry simulation, rely on the formation of a bow shock wave in front of the test model. This, of course, requires that the model be of somewhat smaller dimensions than the gas stream in which it is being tested. If the model is larger than the stream a bow shock will not form and the desired simulation is not achieved [3]. This points out the need for facilities which can accept large test models; and as size increases, mass flow rates must also increase. Larger power supplies are therefore required to supply the additional amounts of energy which must be transmitted to the gas. Larger pumping capabilities are also required to maintain the low pressure within the test tank. All of these factors result in a tremendous increase, not only in the initial investment, but also in operating and maintenance costs. The hyperthermal tunnel test facilities currently in use which are capable of evaluating specimens of appreciable size are nearly all in the one-megawatt and larger power ranges.

Test specimens evaluated in these facilities are more varied than with most of the tests discussed earlier, since the use of models is preferred. These models are often in the form of flat-faced cylinders, hemispheres, paraboloids, wedges, cones or duplicates of actual components, such as sections of leading edges. The method used to analyze the model's performance depends upon the type of material being evaluated but is usually similar to those used in the simpler tests.

5.7. Rocket-Exhaust Simulation

One of the most critical of the wide variety of individual problems involved in the design of a rocket motor is the selection of construction materials, particularly those to be utilized in the combustion and nozzle regions of the engines. Within these regions, the materials are subjected to a high-temperature, high-pressure, chemically-active gas stream traveling at high velocity. The complex interactions of high heat flux, fluid-shear forces, and chemical reactions, in con-

junction with the mechanical forces (pressure and vibration), result in very severe service conditions.

Much of the early materials development effort was directed toward over-
coming the difficulties resulting from the high heat flux and from fluid-shear
and mechanical forces, while the problems of chemical compatibility between
propellant combustion by-products and nozzle materials received only limited
attention. More recently, however, these problems of chemical compatibility have
become more and more serious as advances in fuels technology produce not only
higher combustion temperatures, but also more corrosive exhaust products.

The thermochemical environment associated with a modern rocket engine
presents a very complex picture of transient thermal inputs and incomplete
high-temperature chemical reactions. Analysis by conventional mass- and energy-
transport considerations is virtually impossible except in very simple situations.
Consequently, empirical techniques have been widely employed in the selection
of rocket nozzle materials, and full-scale static rocket firings have become an
accepted means of qualifying final design configurations.

Various techniques are employed in the laboratory evaluation of nozzle materials. Compatibility has been evaluated by heating candidate materials in a static
or slowly moving stream of exhaust constituents. Subscale rocket-engine testing
as typified by the SPAR and MERM engines [122] has been employed quite
extensively and even the arc plasma screening tests have proven of some value.
Each of these techniques, although useful, leaves something to be desired. For
example, the compatibility and arc-plasma screening tests are relatively inexpensive but fall short of simulating the conditions encountered in an actual engine. On the other hand, the subscale rocket motors produce good simulation but
are quite expensive.

When arc plasma jets of megawatt capacity became commercially available,
several manufacturers offered units for testing nozzle-shaped specimens in a
plasma effluent seeded with typical rocket exhaust constituents. These devices
are capable of simulating the rocket exhaust environment very closely since they
are designed to operate at pressures as high as 600 psi. In addition to this, the
plasma facilities offer a distinct advantage over the actual rocket motor tests.
The independent control of the critical parameters, particularly the temperature
and effluent chemistry, is very useful in materials development programs. It
permits studies to be conducted which not only qualify a particular material
for a specific application but also yield useful information on such things as a
material's reactivity to one particular species of the environment.

Smaller arc-plasma facilities in the 50- to 200-kilowatt power range have also
found a degree of popularity. These are usually capable of producing pressures
up to only about 100 psi, however. The small facilities offer the same advantages
as the larger units but are limited to smaller test specimen sizes. They also have
one advantage over the larger units, in that they are many times less expensive
to install and operate. As a result of this and the relatively small amount of
material required to fabricate the small specimens, these small facilities have
been quite useful. They are capable of good simulation of motor conditions, but,
at the same time, permit relatively rapid inexpensive evaluations of a wide variety of experimental materials [123].

The rocket-exhaust simulator usually consists of three major parts; the arc plasma generator, a mixing chamber, and a test section. These three units are secured together so that the plasma gas enters the mixing chamber, where the desired chemical composition is achieved, before it passes through the nozzle test specimen. The mixing section consists of a large plenum chamber with entry ports for chemical additions. Precisely metered quantities of gaseous, liquid or solid phase species are introduced into the plasma effluent in this section. Compatible gases can be premixed in a manifold and are injected radially into the chamber. Liquids are atomized to promote better mixing and are also injected radially. The test chamber usually doubles as a clamping device to secure the specimen. One or more optical ports are provided to permit visual observation of the test as well as to facilitate optical temperature measurement when applicable. Since many of the chemical species which are present in a rocket exhaust are toxic or in the "nuisance" class, some means for collecting and discarding these is usually provided. One such system is the wet-wash scrubber unit [123] which attaches directly to the test section. This unit is designed to cool the gases and neutralize them so that they can be safely discharged into the atmosphere or into a liquid drainage system.

The rocket-exhaust simulation facilities are used to evaluate not only throat-insert materials but also entire nozzles (radiation-cooled or ablative, composite type), motor-case insulation materials and radiation cooled thrust chamber and

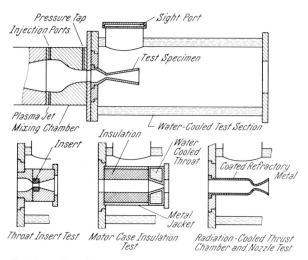

Fig. 56. Various Configurations for Evaluation of Materials in a Rocket-Exhaust Simulator

nozzle skirts as shown in Fig. 56. Entire motor configurations consisting of coated refractory metal thrust chamber, integral nozzle and extension such as are used for altitude control on space vehicles, can also be evaluated.

When evaluating materials to be used in the rocket-motor environment, four major variables existing in the actual rocket-exhaust environment are simulated as closely as is practical. These four variables are: chemical composition and temperature of the exhaust products, chamber pressure and specimen configuration. Chemical composition of the exhaust products is controlled by the injection of additives into the plasma effluent in the same proportions as are present in the rocket engine exhaust. Temperature of the chemically seeded effluent is determined by the input power and efficiency of the plasma unit. Chamber pressure, which is a function of the mass flow rate and nozzle throat diameter, is measured. Test-specimen configurations are usually appropriately scaled models of the actual components.

Through simulation of these factors, tests can be established to determine the behavior of candidate rocket-engine materials and their compatibility with rocket-engine exhausts. As a screening test, large numbers of specimens can be evaluated quickly and inexpensively. Carefully designed full-factorial experiments considering the effects of varying chemistry, temperature, pressure and materials (individually and in combinations) can also be conducted in such a facility.

The propellant systems which are most conveniently simulated are those containing compounds of hydrogen, oxygen, nitrogen and carbon. These systems include the following:

Oxidizer	Fuel
LOX	H_2
(liquid oxygen)	CH_4
	N_2H_4
N_2O_4	MMH (monomethylhydrazine)
	N_2H_4-UDMH (unsymmetrical dimethylhydrazine)
	N_2H_4
MON	MMH
(75 N_2O_4-25 NO)	
H_2O_2	MMH
	N_2H_4

Most recently, systems have been developed which permit the injection of fluorine compounds into the plasma effluent, thereby enabling the simulation of the more advanced propellant systems which utilize fluorine-based oxidizers.

The test procedures are dependent on the specific rocket firing conditions which are being simulated. They are usually characterized by the two general classes of materials:

(1) Those which can be preheated to the desired temperature in an inert environment without serious degradation (this would include most ceramics, graphites, metals and coatings).

(2) Those which depend upon thermally activated physical and chemical changes in the material for thermal protection (these would be primarily ablation and transpiration-cooled materials).

Testing of materials in the first catagory is rather straightforward and simple. Prior to arc ignition, the nozzle or other test specimen is secured in its holder and the entire system is purged for several minutes with an inert gas such as nitrogen or argon. The arc is then ignited and the controls adjusted until the specimen reaches the desired equilibrium temperature. At equilibrium, testing is initiated with the introduction of additives into the plasma effluent. Tests are run either for a predetermined length of time or until specimen failure. The arc is then extinguished, the additive flow cut off, and the specimen allowed to cool in a flow of cold argon or nitrogen.

Cyclic tests of candidate materials for multiple restart rockets follows a similar procedure for the first thermal cycle, that is, the arc is operated on inert gas and no additives are injected until proper test conditions are established. In subsequent cycles, additive injection is initiated at arc ignition and turned off when the arc is extinguished during the cooling cycle. Generally, neither the stabilizing gas flow nor the power-control settings are disturbed during the cooling cycles. These tests are run for either a specified number of cycles or until a failure is produced.

With ablative and transpiration-cooled materials, test levels must be established by destructive testing of trial specimens. At arc ignition, additives are injected immediately, and the plasma torch must be adjusted to operating levels as quickly as possible. Testing is terminated at a specified chamber pressure drop or upon destruction of the specimen.

6. Characterization of Plasma Effluent

The effluent which can be generated by the arc plasma jet is by its very nature well suited to the evaluation of materials in high temperature environments. The various facilities and types of tests which use the plasma for this purpose have been discussed in the previous chapter. All of the tests discussed have a common requirement that the conditions to which the test model is exposed must be well defined.

The parameters which can be used to define the conditions differ from one test to another, as does the accuracy with which the parameters must be determined. In some evaluations it is sufficient to know simply the temperature to which the surface of a specimen is heated and the chemical composition of the gas impinging on the surface. With other evaluations it may be necessary to define, with a high degree of confidence, the enthalpy, gas velocity, pressure, heat flux, chemical composition and degree of non-equilibrium. Not only the bulk or average values of these parameters, but the distribution of each over the entire cross-section of the flow field is often desirable. It is obvious that the degree of sophistication required in the diagnostic instrumentation can be quite high.

As the plasma jet and plasma test facilities have evolved, there was quite naturally a parallel development of diagnostic instrumentation and techniques. Many of the methods were adaptations of existing techniques while others were essentially unique. Many of the techniques developed were designed for a particular or unique application and are not generally applicable to other more conventional problems. The techniques used can be divided into two general categories: those methods of measurement which are carried out on the plasma jet itself, and those instruments or techniques which are used to determine the conditions of the test environment by measurements made directly on the effluent. The latter are discussed in detail in Chapter 7.

A parameter which is of interest in nearly all materials evaluations using the plasma jet is the total or stagnation enthalpy of the test stream. This parameter, along with stagnation pressure, is most often used to define the effluent conditions and can be used to compute the remaining gas properties.

6.1. Energy Balance Technique

The simplest and most widely used method of determining the stagnation enthalpy of a plasma gas stream is to perform an energy balance on the plasma generator [124]. The technique for performing such an energy balance is illustrated schematically in Fig. 57. Both primary arc gas and secondary mixing gas enter the system with an initial enthalpy. Electrical energy is dissipated across the arc by virtue of the resistance which is overcome. A portion of this energy is transmitted to the gas, increasing its enthalpy. Much of the energy from the arc is lost in the form of heat to the electrodes and a portion of the energy from the heated gas is lost to the containing vessel as well. This energy is carried

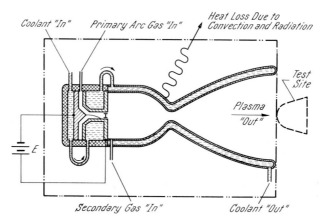

Fig. 57. Envelope for Performing Energy Balance on Plasma Generator

away by the coolant, which enters the system at a low enthalpy level and leaves at a higher level. The energy which remains after all losses are taken into account is the increase in the enthalpy of the plasma gas.

The heat balance is a direct application of the First Law of thermodynamics and says simply that the energy gained by the gas must be the difference between the incoming energy (electrical input) and the total coolant and external losses. This may be stated as follows:

[Energy IN] — [Energy OUT] = [Energy in Plasma]

$$\left[EI\right] - \left[\sum_{i=1}^{n} W_{c_i} C_{p_c} \Delta T_{c_i} + \sum_{j=1}^{p} M_j H_j + Q_{rc}\right] = \left[\sum_{k=1}^{m} W_{g_k} \Delta H_{g_k}\right] \qquad (14)$$

where:

C_{p_c} Specific heat of coolant
E Arc voltage
H_{g_k} Enthalpy of gas
H_j Heat of vaporization of material M_j
I Arc current
M_j Mass loss rate of electrode material
Q_{rc} Energy loss from exterior surface of unit by radiation and convection
T_{c_i} Temperature of coolant
W_{c_i} Mass flow rate of coolant
W_{g_k} Mass flow rate of gas

From Equation 14 it is apparent that the energy content or enthalpy of the emerging plasma effluent for a specific set of operating conditions can be obtained from measurement of the arc voltage and current, the mass-flow rate and temperature rise of the coolant, the mass-flow rate and inlet ambient temperature of the test gas, and the energy loss from the exterior surface of the arc chamber. For all practical purposes, the external surface temperature of the water-cooled plasma arc is not high. Consequently, it will be assumed throughout this discussion that a negligible fraction of energy is lost from the external plasma generator surface by convective or radiative mechanisms. Also the internal loss of electrode or plasma generator materials is generally very slight, and the energy carried away with it is likewise small compared to the input energy.

The energy input term (EI) is to a large degree time-dependent. Fluctuations in the power input can produce errors as large as 50% under certain conditions. The magnitude of the error will depend on the amplitude of the unsteady compared with the steady portion of the current and voltage and also on the instantaneous phase relationship between current and voltage. The power input portion term should be written:

$$EI = \frac{1}{t} \int_{o}^{t} EI \, dt \qquad (15)$$

As a consequence, each plasma generator should make use of voltage-current traces from an oscilloscope during its development in order to ascertain the variations of the voltage-current input. If these traces show significant unsteadiness it is recommended that additional methods of input power measurements be pursued.

In order to measure power directly, a wattmeter can be employed [125]. As a precaution in the use of the wattmeter, reversed polarity readings are used. For those plasma generator facilities which operate under known and steady input power, the use of a voltmeter and ammeter is recommended because of their high degree of accuracy. Suitable instruments for such voltage measurements have been described by the Instrument Society of America [126]. The measurement technique to be used can be either a voltage-divider network or a direct-reading instrument. It is highly desirable to be able to record the voltage such that time variations are a part of the test data. Accuracy of the voltage reading within $\pm 1\%$ is generally recommended. The voltage measurement should be taken at the electrode terminals of the plasma-generator circuit.

The measurement of plasma arc current can be accomplished with an ammeter equipped with a precision shunt and these readings should also be within $\pm 1\%$. The Instrument Society of America [126] has compiled a list of equipment suitable for measuring arc current. If a precision shunt is utilized, the temperature across the shunt must be constant and within the stated limits given by the manufacturer. Arc current should be measured as close to the plasma arc terminals as feasible. If cooled electrical power leads are used, as is the case with many of the smaller plasma units, these leads generally act as inlet and outlet passages for coolant transfer to and from the plasma jet. In these cases the voltage and current measurements should be made at a point which includes losses in these leads. In this configuration the I^2R losses in the leads are absorbed in the form of heat by the cooling fluid.

The *coolant flow rate* to each cooled component of the plasma facility must be measured. An error in these measurements of not more than $\pm 2\%$ is acceptable. Suitable equipment which can be used [126] includes turbine flow meters, head flow meters, area flow meters, etc. Care must be exercised in the use of these devices. In particular, it is recommended that appropriate filters be placed in all water inlet lines to prevent particles or unnecessary deposits from being carried to the cooling passages, pipe and meter walls. Flow rates must be properly adjusted so that bubbles and cavitation are eliminated and that water vapor formation does not occur.

The method of *temperature measurement* must be sufficiently sensitive and reliable to insure accurate measurement of coolant temperature rise. Procedures similar to those recommended by ASTM [127] should be adhered to in the calibration preparation of temperatures sensors. The bulk or average temperature of the coolant must be measured at the inlet and outlet lines of each cooled unit. The error in measurement should be not more than $\pm 1\%$. The temperature-indicating devices should be placed as close as practical to the plasma jet in the inlet and outlet lines and it is generally recommended that no additional apparatus be placed between the temperature sensor and the plasma. If at all possible, the temperature measurements should be recorded continuously. A variety of commercially available temperature sensors are capable of continous recording [126]. During operation of the plasma, care should be taken to minimize deposits on the sensors and to eliminate any possibility of sensor heating because of test-sample radiation to the sensor. In addition, all coolant lines should be shielded from direct radiation from the test sample.

The *flow rate* of each gas entering the plasma generator must be measured with an error no greater than $\pm 1.5\%$. There are many suitable commercially available gas-flow meters [126], but for most applications the orifice or rotameter type will suffice. Pressure measurement should follow the manufacturer's instructions and calibration charts. The flow meters should be placed in that portion of the gas lines where disturbances are at a minimum. Since the incoming-gas temperature will not differ substantially from room temperature, it will usually have negligible effects on the calculation of total enthalpy. However, if it is practical, a thermal sensor may be installed in an inlet section where the gas flow is free from disturbances. It is desirable to measure temperature of the inlet gas stream in a settling or plenum chamber where the gas velocity has been minimized. The output of the thermal sensor device should be recorded continuously.

Since the energy balance requires that the plasma generator operate as a steady-state device, all calculations must be made using only measurements taken after it has been established that the device has achieved steady operating levels. To assure steady flow or operating conditions, the above mentioned parameters should be continuously recorded so that instantaneous measurements are available to establish a measure of steady-state operation. It is highly desirable that whenever possible separate measurements be made of the desired parameters.

In all cases, the parameters of interest, such as arc voltage, arc current, gas and coolant flow rates, and cooling-fluid temperature rises, should be automatically recorded throughout the calibration period. Recording speed will depend on the variations of the parameters being recorded. Generally, a recorder response time of one second or less for full scale deflection is desirable.

Although the procedure to be followed in obtaining the data required to calculate enthalpy is not complicated, certain precautions must be taken. It is essential, for instance, that the entire system operate at steady-state conditions prior to, and during, the time when data are taken. Therefore the cooling system must be operating at steady-state conditions prior to arc initiation. Gas inlet and cooling-fluid inlet and outlet temperatures should be recorded throughout the calibration period. Steady-state operation assumes, of course, that the coolant temperature, arc current and voltage, and gas and coolant flow rates are steady and not changing with time. In particular, the coolant flow rate should not change during arc operation. After the arc is extinguished, the gas inlet and coolant temperatures and flow rates should again be checked to allow comparisons to be made with the values obtained prior to arc ignition. Any changes which occur between pre- and post-test values may indicate a buildup of deposits in the cooling system or some other difficulty which could alter the results of the energy balance.

Because of internal friction there is generally a small but significant rise in the temperature of the coolant, even when the plasma is not operating. It is assumed that these same conditions exist while the arc is in operation. The ΔT_c (temperature rise of the coolant) in Equation 14 is therefore not taken as the difference between coolant inlet and outlet temperatures but as the difference between temperature rise with the arc operating and temperature rise without the arc operating. All temperature measurements used in this heat balance are bulk or average temperatures. By eliminating the insignificant terms and rearranging, Equation 14 may be rewritten to define the total enthalpy of the emerging plasma as follows:

$$H_{f_g} = \frac{\displaystyle\sum_{k=1}^{m} W_{g_k} H_{g_k}}{\displaystyle\sum_{k=1}^{m} W_{g_k}} = \frac{\displaystyle\sum_{k=1}^{m} W_{g_k} H_{og_k} + EI - \sum_{i=1}^{n} W_{c_i} C_{p_c} (\Delta T_f - \Delta T_o)_{c_i}}{\displaystyle\sum_{k=1}^{m} W_{g_k}} \tag{16}$$

where:
H_{f_g} Total enthalpy of plasma
H_{og_k} Enthalpy of each incoming gas
ΔT_f Temperature rise in coolant with arc operating
ΔT_o Temperature rise in coolant without arc operating,

the remainder of the quantities being defined as before. H_{f_g}, as defined by Equation 16, is a measure of the average total enthalpy of the plasma gas as it passes through the exit plane of the plasma-generating system.

One of the limitations of this method of determining the enthalpy is that the value obtained is the average value over the cross section of the flow field. Since the nozzle walls, and other containment parts, are always cool relative to the plasma effluent, a cool boundary layer surrounds the hottest portion of the flame. This of course means that the enthalpy of the hot center "core" of the gas stream is higher than that of the cooler outer edges. The severity of this coring varies not only from unit to unit but also as a function of enthalpy or power level, condition of electrodes and mass-flow rates, among other things. The main difficulty arising from this coring is that the test specimen normally is exposed to only a fraction of the total flame cross section and this is usually in the center, higher enthalpy region.

A further disadvantage in this method is that the measured enthalpy represents the value which exists as the effluent passes through the exit plane of the unit. The enthalpy generally drops off quite rapidly as the distance from this exit plane becomes larger. Unless the test specimen is placed precisely at this plane, it will not be exposed to the conditions as determined from the heat balance method. The degree of variation in enthalpy of the gas after it leaves the generating unit differs from unit to unit and is unquestionably most severe in a facility which operates without a closed low-pressure test tank. In these cases a large volume of air from the surrounding atmosphere is entrained into the plasma stream. Within a very short distance (slightly more than one inch) the flow stream can be made up of from $1/3$ to $1/2$ entrained room air [120]. The low-pressure area which is formed by the high-velocity gas stream or jet causes the air to be inducted into the test stream. This room-temperature air mixes with the hot effluent and lowers its average enthalpy by a substantial amount.

The heat-balance method for determining total enthalpy has one advantage which accounts for its widespread use; that is, the enthalpy can be determined not only before or after a materials test is conducted but the measurements can (and usually are) made while the actual test is in progress. The heat balance is generally used for each individual test and even in cases where the enthalpy is not used directly in determining materials performance (such as in splash-type screening or dynamic-oxidation tests) it is used as an indication of the performance of the plasma jet.

The efficiency of a plasma jet is defined as the percentage of input power which is contained in the plasma as it leaves the unit.

$$\text{Efficiency} = \frac{H_{f_g} \sum_{k=1}^{m} W_{g_k}}{EI} \times 100 \tag{17}$$

Overall plasma generator efficiencies vary from less than 10% to greater than 70%. Some of the factors which can effect efficiency are basic plasma jet design, the size of mixing chambers, plenums and nozzles which are cooled, the amount of secondary gases which are added, and the operating conditions. In particular, any increase in pressure within the plasma jet results in much larger quantities of energy being transmitted to the cooled walls and thus reduces the efficiency drastically.

Although this method of determining the total enthalpy is direct and straightforward, some precautions must be taken. For example, extreme care must be exercised when making measurements during a material test to minimize any effects which the presence of the test model may have on the system. If the test model is placed very near the exit nozzle, as shown in Fig. 57, the model will reradiate energy to the interior of the nozzle. The extent of this reradiation is a function of model material and relative size. A large test model will produce a more serious effect than a smaller one, and a material such as graphite, whose temperature stabilizes at a very high level and relies on reradiation as a major method of cooling, will also present a serious effect. By not accounting for this reradiation during testing, the energy balances of the whole system can be altered and thus affect the enthalpy determination as derived using the energy-balance technique.

Because the efficiencies of many plasma generators are low (10% or less) the calculated values of enthalpy or energy in the gas depends upon the difference between two large, measured quantities of nearly equal magnitude. Therefore, the accuracy of the measurement of the primary variables must be high, all energy losses must be correctly taken into account and steady-state conditions must exist both in arc performance and fluid flow.

6.2. Sonic-Flow Method

Another technique which has been devised to determine the average total enthalpy of a plasma stream involves measuring the operating conditions of the arc jet. This method, which is termed the sonic-flow method, is based on the relation between the enthalpy of a gas in a reservoir and the rate at which the

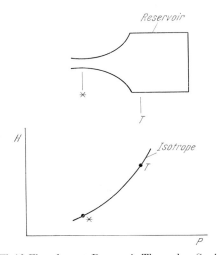

Fig. 58. Fluid Flow from a Reservoir Through a Sonic Nozzle

gas can flow through a nozzle under choked-flow, equilibrium, isentropic conditions [128]. Since the flow rate through a nozzle under these conditions is determined by the pressure and the total enthalpy of the gas in the reservoir, then, conversely, measuring the gas flow rate and the pressure in the reservoir will permit a computation of the total enthalpy of the effluent discharging at sonic velocity through the nozzle.

The relationship required is readily derived for a perfect gas such as air at near atmospheric conditions. The expansion of the gas from the mixing or plenum chamber, which is considered the reservoir, through a nozzle is depicted in Fig. 58. To derive the desired relationship the following assumptions must be made:

(1) The flow is in thermal equilibrium.
(2) The flow is isentropic.
(3) The flow is one-dimensional and homogeneous.
(4) The flow is steady with time.

The equations which govern the flow from a reservoir through a nozzle are the Equations of:

$$\text{Continuity} \qquad w = \varrho V A \qquad\qquad (18)$$

$$\text{Energy} \qquad h_t = h + V^2/2gJ \qquad\qquad (19)$$

$$\text{State} \qquad \varrho = \frac{P}{(R_o/m)T} \qquad\qquad (20)$$

and the equation describing the flow process, which is isentropic $(ds = 0)$:

$$\frac{dp}{p} = \gamma \frac{d\varrho}{\varrho} \qquad\qquad (21)$$

In these relationships the following notation is used:

A Cross sectional area
g Gravitational acceleration
h Enthalpy
m Molecular weight
p Pressure
R_o Universal gas constant
T Absolute temperature
V Velocity
w Mass flow rate
γ Ratio of specific heats
ϱ Density
t Stagnation state
$*$ Sonic value at nozzle throat

When choked-flow conditions exist, the velocity is sonic at the throat and the mass velocity is:

$$\frac{w}{A} = \varrho^* V^* \qquad\qquad (22)$$

This can be expressed in terms of conditions within the reservoir by making use of Equations 18, 19, and 20.

$$\frac{w}{A^* p_t} = \frac{\varrho^* \sqrt{2gJ(h_t - h^*)}}{\varrho_t (R_o/m) T_t}$$

$$\frac{w}{A^* p_t} = \left(\frac{\sqrt{gJ}}{R_o/m}\right)\left(\frac{\sqrt{h_t}}{T_t}\right)\left\{\frac{\varrho^*}{\varrho_t}\left[2\left(1 - \frac{h^*}{h_t}\right)\right]^{1/2}\right\} \qquad\qquad (23)$$

This expression is applicable to either perfect-gas or real-gas flow. For a perfect gas such as low-temperature air, and $\gamma = 1.4$, Equation 23 reduces to the much simpler form:

$$\frac{w}{A^* p_t} = \frac{551}{\sqrt{h_t}} \tag{24}$$

For imperfect gases such as air at higher temperatures, other effects must be considered. The specific heat ratio (γ) is dependent on the temperature, and as enthalpy levels of 1000 Btu/lb and greater are achieved, the effects of gas molecule dissociation and atom ionization must be taken into account. Because of dissociation, both the molecular weight of the mixture and the specific-heat ratio are dependent not only on temperature but also on pressure. Consequently, no simple expression for these terms can be written. Also, additional equations, relating the dissociation phenomenon to the state of the gas, are required to solve the sonic-flow relation analytically. Because of these complications which arise for air at elevated temperatures, the method for obtaining the sonic-flow relation which applies to plasma jets was derived from empirical considerations. The technique consisted of obtaining the mass velocity at the sonic point as given by Equation 22. Successive approximations were made involving Equations 18—21 and an Equation of State which takes the effects of dissociation, etc. into account [129]. The calculations were made for pressures from $1/4$ to 100 atmospheres and for enthalpies from 1000 to 10,000 Btu per lb. The results of all these calculations revealed that for all pressures, the value of $w/A^* p_t$ fell within 4% of a mean curve. The equation of this curve was therefore taken as the solution for the sonic-flow expression for high-temperature air as a real gas. The least-squares fit to the data is given by:

$$\frac{w}{A^* p_t} = \frac{280}{(h_t)^{0.397}} \tag{25}$$

The sonic-flow relationships for three enthalpy levels are shown in Fig. 59. The region up to 250 Btu per lb shows the relationship for a perfect gas (Equation 24). From 250 to 1000 Btu per lb the curve takes into account the effects of variable specific-heat ratio but not dissociation. The upper portion of the curve, from 1000 to 10,000 Btu per lb represents Equation 25, which applies for air as a real gas. This overall curve yields the total enthalpy of the emerging stream in terms of rather easily measured quantities, i.e., mass-flow rate, nozzle throat area and the reservoir or chamber pressure.

The small pressure dependency which does exist in the real-gas region of this curve can be handled by application of a special test technique along with the sonic-flow formula. This technique involves measuring the necessary parameters before the arc is ignited (but after constant flow has been established) and then measuring the pressure rise as the arc is initiated. A family of curves relating enthalpy to pressure rise is available for various operating pressures. These curves were constructed from point calculations rather than from a relationship such as Equation 25. By using these curves, the small pressure dependency can be taken into account by interpolating between the curves of this family. In many instances, however, the curve of Fig. 59 is sufficiently accurate and the use of the pressure-rise technique is not necessary.

The use of the sonic-flow method for determining total enthalpy is not without limitations. The most obvious of these is that it can be used only on plasma units which operate in the sonic or supersonic region. Also, for this method to be applicable, the energy must be imparted to the gas in an essentially stagnant

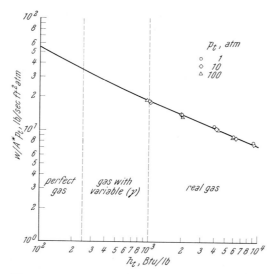

Fig. 59. Equilibrium Sonic-Flow Formulation for Air
(From WINOVICH [128])

region. This would rule out, for example, devices where the arc occurs through the nozzle and adds significant energy in either the inlet or supersonic portion of the nozzle.

The assumption of equilibrium nozzle flow which underlies the formulation above has been critically reviewed [128]. On the basis of the transit time of the flow to the throats of typical nozzles, and vibrational and dissociative relaxation times, the flows will be in vibrational equilibrium for practical ranges of reservoir pressures, and in dissociative equilibrium for reservoir pressures above about 10 atmospheres. For the nozzle-inlet flows which will be chemically frozen for pressures below 10 atmospheres, the total enthalpy deduced from equilibrium sonic-flow relations will be low by about 13% at 10,000 Btu per lb.

Consideration was also given to errors in the formulation arising from the effects of irreversibility, heat transfer, and boundary layer. Effects of irreversibility (principally a reduction in enthalpy due to nozzle-inlet heating) are shown to be small for typical inlet designs, even though local heating rates may be high. It has been demonstrated that the heat-transfer and boundary-layer effects yield compensating corrections to the total enthalpy deduced from relations for equilibrium sonic flow. This result was shown to be due to the nozzle-discharge coefficient approaching unity for the small ratios of wall temperature to free-stream temperature that always occur in the nozzle inlets of arc heaters.

Values of total enthalpy, determined for one particular plasma jet by use of the equilibrium sonic-flow method, were compared with corresponding values deduced from a combination of measurements of stagnation heating rates and pressure, and an appropriate heating-rate theory. The comparison indicated that when the necessary assumptions of the sonic-flow method are not seriously violated, the enthalpy values predicted agree with those determined by other methods.

7. Diagnostic Techniques

All properties of a flowing effluent determined from measurements made on the plasma generator itself are average values. These measurements are limited in that they yield values of the parameters which exist as the gas passes the exit plane of the generator. To determine gas properties as they exist at the test site or at specific locations in the cross section of the flow field requires measurement of parameters within the stream itself.

7.1. Enthalpy Probes

A map or profile of the enthalpy which a model encounters during test is nearly always desired. These data can be most valuable in determining the performance of the test material. There have been a number of total-enthalpy probes

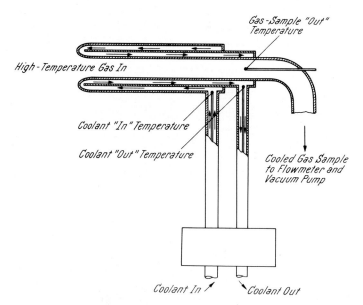

Fig. 60. Simple Calorimetric Enthalpy Probe

developed over the past decade which were designed to measure enthalpy at a precise point within the flame rather than over the cross section. The majority of these probes fall into a category which is described as calorimetric in nature.

These calorimetric probes measure, in one manner or another, the energy content of a small sample of gas which has been extracted from the plasma stream. The calorimetric enthalpy probes simply cool a metered quantity of the high-temperature gas stream to a state where more conventional means for determining enthalpy can be utilized. The quantity of energy which is removed from the gas in bringing it to the lower temperature is also measured. A diagram of a simple calorimetric enthalpy sensor is shown in Fig. 60.

These probes have also been called conservation-of-energy probes since they represent an application of the First Law. Since there is no appreciable work added or removed from the system, the energy which is extracted from the gas which passes through the sensor can be equated to the energy increase of the coolant as it passes through the instrument. The energy balance can thus be written:

$$W_g \left(H_{g_o} - H_{g_f} \right) = W_c C_{p_c} \left(T_{c_f} - T_{c_o} \right) \tag{26}$$

where:

C_{p_c} Specific heat of coolant
H_{g_o} Enthalpy of gas entering sensor
H_{g_f} Enthalpy of gas exiting sensor
T_{c_o} Coolant temperature entering sensor
T_{c_f} Coolant temperature exiting sensor
W_c Mass-flow rate of coolant
W_g Mass-flow rate of gas through sensor

This can be rearranged to yield an explicit expression for the initial total enthalpy of the gas stream as it enters the system:

$$H_{g_o} = \frac{W_c}{W_g} C_{p_c} \left(T_{c_f} - T_{c_o} \right) + H_{g_f} \tag{27}$$

As was stated earlier, the calorimetric probe is designed to cool the gas sample which passes through it to a relatively low temperature. The stagnation enthalpy of the gas as it leaves the sensor (H_{g_f}) is then well within the range where it is represented by a linear function of stagnation temperature:

$$H_{g_f} = C_{p_g} T_{g_f} \tag{28}$$

where the temperature is measured relative to the temperature at which the reference or zero enthalpy has been established. As the total enthalpy of the gas entering the sensor becomes very large, the last term in Equation 27, as defined by Equation 28, becomes negligible.

Calorimetric-type total-enthalpy probes do not distinguish between the different forms in which the energy may be present in the stream, nor do they depend upon a knowledge of the state of the gas to yield valid data. For that matter they do not require a knowledge of the composition of the incoming fluid, unless the term represented by Equation 28 for the exit enthalpy is of sufficient relative magnitude to warrant consideration. Energies of dissociation or ionization can be recovered in a properly designed sensing probe.

The simple probe which is depicted in Fig. 60 shows the basic concept around which most calorimetric enthalpy sensors are designed. It is fairly obvious, however, that such a simple design could be used to make reliable measurements in only very specialized applications. It is limited, in fact, to flow conditions where the entire flow stream is swallowed by the probe. If this were not the case, the probe would be heated from the exterior surfaces as well as along the interior. This increase in energy absorbed by the coolant would produce erroneous, high values of measured enthalpy. One of the principal reasons for using a probe rather than the energy balance method or some other indirect measuring technique is that the enthalpy varies across the cross section of the gas stream. It is therefore desirable to keep the size of the probe small, to permit measurements to be made at precise locations within the stream. A probe which requires the entire stream to pass through its interior obviously cannot meet this requirement. It is also desirable to limit the overall size of the probe to minimize the disturbance of the flow field which results from the very presence of a probe.

In order to make use of the basic concept of the simple calorimetric probe, many variations and alternate designs have come into existence. One of the least complicated variations involves substituting an adiabatic wall between the two channels of coolant flow and moving one of the thermocouples to the leading or gas-inlet end of the coolant channel [130]. This design reduces the effects of heating from the external surfaces. Although the coolant is heated as it passes through the outer channel, the initial coolant temperature (T_{c_o} in Equation 27) is determined by the thermocouple placed at the point where the coolant passes from the outer channel into the inner channel. A true adiabatic wall between the incoming (outer) and outgoing (inner) coolant channels eliminates heat transfer through this wall between the points where the coolant temperature measurements are made. Obtaining such truly adiabatic conditions across this wall, which must be kept very thin to maintain small overall dimensions, has been found to be the major draw-back of this particular type of sensor.

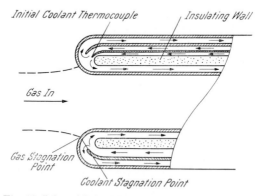

Fig. 61. Inlet of Split-Flow Type of Enthalpy Probe

The *split-flow probe* represents another variation of the basic enthalpy sensor system. In this probe, which is shown schematically in Fig. 61, an additional tube is placed concentrically between the outer wall and the insulating divider wall. The incoming coolant passes though the central channel formed by the adiabatic tube and the newly added tube. As it reaches the tip of the probe, the coolant divides into two streams, one passing along the outer channel to cool the instrument, and the other passing along the inner wall to cool the gas sample being analyzed. A thermocouple to determine the initial temperature of the coolant is placed at the point where the flow splits. The remainder of the probe is similar to the simple probe described earlier.

Earlier models of the split-flow probe did not have a thermocouple placed near the tip, but rather measured the initial coolant temperature at a point near entry into the instrument. Placement of the thermocouple as shown in Fig. 61 minimizes the effect of heat transfer from the inner jacket to the inlet coolant, and almost entirely eliminates the effect of outer-jacket heat transfer to the inlet flow. These earlier probes had exhibited serious limitations in sensitivity to coolant flow variations and in heat transfer capability.

The split-flow probes are sensitive to variations in gas flow conditions. An accurate measure of heat flux from the gas sample can be realized only if the stagnation point of the incoming gas coincides with the stagnation point of the coolant flow. If this condition does not exist, the coolant will indicate the energy content of either a larger or smaller flow tube of hot gas than is actually passing through the probe. The stagnation point of the coolant is essentially fixed, although it can be shifted slightly by controlling the flow through the inner jacket relative to that through the outer passage. On the other hand the stagnation point of the incoming gas varies appreciably as a function of pressure, Mach number, flow rate through the sensor, whether the normal shock wave is swallowed or if a detached bow shock is formed, etc. Shifting of the gas stagnation point away from direct coincidence with the coolant stagnation point can affect the measured enthalpy from 5—10% about a mean calibration [130]. These values, which represent a total variance of up to 20%, are typical over a simulated-reentry trajectory from orbital velocities.

Generally speaking, this stagnation-point shifting error is decreased by reducing the radius of curvature of the probe inlet leading edge. Sharpness of this leading edge is limited, however, since the heating rate at the stagnation point is inversely proportional to the square root of the leading edge radius. A correction factor can be applied in the case of a well-calibrated probe since the effect is systematic. Another alternative is to adjust the flow rate of gas through the sensor to keep the gas-flow and coolant-flow stagnation points coincident. The technique used in operating the split-flow probe is identical to that for the simple unit and the enthalpy of the unknown incoming gas is given by Equation 27. Caution must be exercised, however, to use the value of coolant flow rate (W_c) of only that portion of the total flow which passes through the inner jacket.

The split-flow type of enthalpy probe offers several advantages over the simple probe. The coolant temperatures (initial and final) to be used in calculating the amount of energy removed from the gas sample can be more accurately determined with the split-flow concept. The coolant flow rate through the inner and outer jackets can be independently adjusted. This means that the coolant temperature rise in the inner jacket can be maintained at a high value for greater accuracy and also that some control is maintained over the amount of energy which is extracted from the gas sample as it passes through the instrument. Coolant flow through the outer jacket can be regulated to meet the external surface cooling requirement. Another advantage of this system is that coolant is supplied to the stagnation region of the probe, where critical heating occurs, at its initial (lowest) temperature.

The most widely used method for determining the enthalpy at precise locations within a plasma stream is the unique *tare technique* [131]. This technique can be used with any of the probe designs discussed, i.e., simple, modified or split-flows types. A major problem which exists with all probes when they are miniaturized is the error which is introduced by heating from the gas which passes over the external surfaces of the instrument. The tare-measurement effectively eliminates this error.

In using the tare measurement technique a valve in the gas sample line is first closed, thus preventing gas from entering the interior of the probe. During this mode of operation, all coolant-temperature and flow-rate information is recorded. Since there is no gas flowing through the instrument, the heating which occurs during this time is obviously a result of external heating. The valve is then opened and the gas to be sampled is permitted to flow through the probe as well as across the outside surfaces. The measurements are then taken as previously described, while the probe is in this operating position. The rate of energy removal from the gas sample is obtained by subtracting the first value (external heating only) from the second value (both external and internal heating). This is why the technique is called the "tare" method. Slight modifications to Equation 27 result in the working equation for use with this technique:

$$H_{g_o} = \frac{W_c}{W_g} C_{p_c} [(T_{c_f} - T_{c_o}) \text{ flow} - (T_{c_f} - T_{c_o}) \text{ no flow}] + H_{g_f} \qquad (29)$$

The use of Equation 29 is dependent on the assumption that the coolant flow rate does not vary between the two modes of operation, "flow" and "no flow".

This tare technique is obviously dependent on the assumption that the overall heat transfer to the exterior of the probe is identical under flow and no-flow conditions [132]. The "tare" measurement principle is valid only if the sample-gas flow rate is sufficiently low, and the probe-tip geometry designed such that the flow pattern does not change between the two measurements. This immediately points up a second limiting feature, that of probe sensitivity. As the gas sample flow rate is reduced to zero, which represents the "ideal" condition for tip heat-transfer duplication, the measured change in energy extraction also drops to zero. Thus there is an optimum value of gas-sample flow rate which will be sufficiently high to provide adequate sensitivity but sufficiently low so as to not introduce errors due to non-duplication of the tip flow conditions. The probe sensitivity of the split-flow type of probe is considerably better than with the simple or standard type.

The geometry and relative dimensions of the probe tip have a significant effect on the duplication of external heating conditions from one mode of operation to the other, even at low subsonic flows. An open, wide-mouth inlet configuration, as shown in Figs. 60 and 61, for example, would be highly sensitive to gas sample-flow rate, since the stagnation point position would vary considerably as the gas-sample flow rate changes. This produces wide variations in measured enthalpy because of the extreme sensitivity of this type of probe to the high heat-transfer rate at the stagnation-point. A more acceptable geometry would be a hemispherical or even a flat-faced probe with a small, nearly sharp-edged gas-sample aperture. This geometry, which is necessary for supersonic stagnation-point determinations, also minimizes stagnation-point displacement with changing gas-sample flows, and therefore minimizes sensitivity to gas-sample flow rates.

Another sensor design which is intended to reduce the effects of external heating is the *aspirating probe*. The operating principles for this probe are identical to the other calorimetric instruments discussed earlier. The only major difference is that they are constructed of two cooling jackets which operate independently from each other. The inner jacket, which is used to measure the quantity of energy removed from the gas sample, is separated from the outer jacket, which protects the probe, by an air gap. The "tare" method of operations is generally not required for these probes since a properly designed air gap serves as very effective insulation.

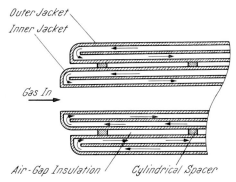

Fig. 62. Inlet of Blunt-Face Aspirating Enthalpy Probe

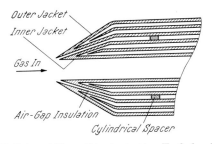

Fig. 63. Inlet of Sharp-Lip Aspirating Enthalpy Probe

Two general models for this probe are shown in Figs. 62 and 63. The blunt-face model depicted in Fig. 62 is for stagnation-point measurements and the sharp-lipped conical-inlet model of Fig. 63 is for supersonic (shock-swallowing) applications [133].

The principal disadvantage of the blunt-face design is the difficulty involved in defining the dividing streamline between the sample flow entering the probe and the diverted flow which passes around it. This dividing streamline should terminate precisely at the air gap, so that all gas cooled by the blunt inner calorimeter face becomes part of the sampled flow. It is possible to define the location of the dividing streamline by running a series of measurements at the same incoming gas enthalpy but at different probe gas-sample flow rates, thereby moving the stagnation point location from the centerline (which would occur at zero gas-sample flow) toward the outer probe periphery (at maximum gas-sample flow). By plotting measured enthalpy against gas-sample flow rate the desired sample flow can be determined from the flat plateau of the resulting plot where enthalpy is essentially independent of sample-gas flow rate. It must be pointed out however, that the optimum gas-sample flow rate thus determined would be valid only for those free-stream flow conditions. Since the pattern of flow over the probe tip is a function of gas velocity, pressure, etc., this method of locating the stagnation point must be repeated whenever the free-stream conditions are changed. This can be very inconvenient.

If the free-stream (static) value of enthalpy is desired rather than the stagnation value, or when it is either impractical or impossible to perform adequate dividing-streamline location tests, the shock-swallowing model of Fig. 63 must be used. This type of design represents the "ideal" in enthalpy probes for supersonic or hypersonic flows but, as is the case with most "ideal" devices, they are extremely difficult to realize in the form of practical hardware. Even when these probes are built with large outside diameters (over $3/4$ inch) it is difficult to maintain a sharp lip with the required number of separate tubing junctures and sufficiently thin walls to support the necessary heat-transfer rates. Considerable effort is being expended however, on the development of a suitable small, reliable, sharp-lipped probe.

One of the principal sources of difficulty in all calorimetric type sensors is the heat-transfer capability of the probes. Although heat-transfer capability varies considerably from application to application and from probe to probe "limiting values" of gas-stream energy density of from 3000 to over 20,000 Btu per ft² sec have been cited [131]. The heat-transfer capability depends not only on flow-energy density, but also on the nature of the gas, its transport properties and the pressure-enthalpy balance for given energy density, as well as the detailed geometry of probe design. Often the "limiting value" is dictated not by the stagnation-point heat-transfer load, but by the size of the jet and flow field being studied. For example, a cooled probe immersed only a short radial distance into a stream of very high enthalpy gas may operate satisfactorily, whereas the same probe immersed for a greater distance into a stream of relatively low enthalpy could conceivably fail due to the excessive total energy which the coolant must absorb in order to cool the immersed portion of the probe. Because of the desire to maintain the overall outside dimensions of a probe as small as possible, coupled with the necessity to provide flow rates of coolant adequate to protect the instrument from damage, very high pressures, on the order of one thousand psi or above, are required to operate these sensors.

Another calorimetric enthalpy probe is the *transpiration-cooled, evaporating-film type* of instrument [134]. This type of sensor, which is a diluent type of probe, operates on the same basic technique as those described previously, i.e., cooling a gas sample to the point where direct measurement can be made and performing an energy balance on the heat exchanger which cooled the gas. The major difference in this type of sensor is that the gas is cooled by an evaporating liquid film. A general schematic of a typical evaporating-film unit is shown in Fig. 64. The film of liquid coolant which is used to cool the incoming gas stream by evaporation into the gas stream is introduced near the inlet of the probe. At the rear of the probe, the temperature of the gas mixture (gas sample plus coolant vapor) is sufficiently low that the temperature may be measured by conventional

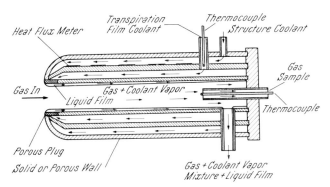

Fig. 64. Evaporating-Film Enthalpy Probe

means. For purposes of the energy balance, a sample of the gas mixture is taken at this point also and this sample is analyzed to determine its composition. The bulk of the gas mixture and remaining liquid layer are drawn from the probe through another tube. The external surfaces of the instrument are generally also protected from heating from the surrounding high-temperature environment by transpiration cooling.

The energy balance for this system can be written as follows:

$$W_{g_o} H_{g_o} + W_{c_o} H_{c_o} = W_{m_f} H_{m_f} + W_{c_f} H_{c_f} \pm \int_0^L q(x)\, dx \tag{30}$$

where the subscripts g, c, and m indicate gas, coolant and mixture, respectively, and o and f incoming and final conditions. The last term in this expression represents the heat transfer across the wall dividing the transpiration film coolant and the coolant structure. This rate of transfer is determined from the heat-flux gauges attached to the dividing wall. By using the Continuity Equation to account for the mass-flow rates and by defining the mass fractions y as:

$$y_{c_f} = \frac{W_{c_f}}{W_{m_f}} \tag{31}$$

and

$$y_{g_f} = (1 - y_{c_f}) = \frac{W_{g_o}}{W_{m_f}} \tag{32}$$

the operating equation can be written. This equation, which can be used without regard to the liquid coolant being evaporated, so long as enthalpy data on the coolant are available, is:

$$H_{g_o} = \frac{y_{c_f}}{1 - y_{c_f}} \left(H_{c_{f, vap.}} - H_{c_o} \right) + \left(\frac{W_{c_o}}{W_{g_o}} - \frac{y_{c_f}}{1 - y_{c_f}} \right) \left(H_{c_{f, liq}} - H_{c_o} \right)$$

$$\pm \left(\frac{1}{W_{g_o}} \int_o^L q(x) \mathrm{d}x \right) + H_{g_f} \tag{33}$$

Here the added coolant subscripts *vap* and *liq* are used to designate the energy involved in vaporization of the coolant and that which is absorbed by the portion of the coolant which remains liquid.

For the maximum range of operation, the coolant which is utilized should have a high heat of vaporization and a high heat capacity in the gaseous phase. Water is used almost exclusively since it not only meets these requirements but is also quite easy to work with.

For determination of gas enthalpy by this technique, the following measurements need to be taken:

- Temperature of the exit gas mixture,
- Vapor composition of the exit gas mixture,
- Gas mixture flow rate,
- Transpiration coolant flow rate.

With the exception of the exiting mixture composition these measurements are relatively straightforward. According to its developers, one of the major advantages of the evaporating-film probe over other, more popular techniques, is that it can be operated continuously. A technique is therefore needed which will permit continuous measurement of the water vapor content of the exiting mixture.

Water vapor content of gas streams can be determined by using heat of absorption, thermal conductivity, infrared techniques, or gas chromatography. Each of these four techniques has basic shortcomings which negates its use in conjunction with the evaporation-film enthalpy probe. A problem common to all is the maximum operating temperature, which is on the order of 120° F for most commercially available units. The heat-of-absorption techniques are further limited to concentrations on the order of three mole per cent. Extension to higher concentrations is prohibited by the rapid decay of the absorbing material due to the large amount of water present. The thermal conductivity instruments have slow response times and rather poor accuracy for this application. Instruments based on infrared measurement have moderately fast response times but are limited to concentration levels of about 10% water vapor. The gas chromatograph techniques are essentially noncontinuous because the gas can be sampled only at intervals.

An instrument which uses the dielectric constant of the mixture as a quantitative tool was developed [134] for use in conjunction with this enthalpy sensor. This instrument makes use of an electronic system capable of continuous recording of the capacitance of a specifically designed capacitor. Although problems were encountered with the long-term stability of this system it has been reported that an accuracy of $\pm 2\%$ has been realized and that measurements can be made over the range of 10 to 90 mole per cent water vapor at temperatures up to 300° F and with a response time of approximately one second.

In another variation of the transpiration-cooled, diluent-type of total-enthalpy probe, shown in Fig. 65, a gaseous-phase coolant is injected directly into the hot gas stream inside the probe. The injected gas mixes with the test fluid resulting in a final gas mixture which is considerably diluted. The gas mixture is sampled and its composition determined by the use of a gas analyzer. Unlike the liquid-film evaporation unit, all of the diluting gas is assumed to be mixed uniformly with the test gas. It is therefore not necessary to measure the mass-flow rates of coolant into and out of the probe. By measuring the total flow rate exiting

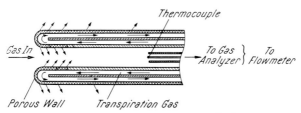

Fig. 65. Diluent-Type Enthalpy Probe

from the probe and the percentage of this flow which is diluent, it is possible to calculate the flow rates of diluent and principal incoming gas. The enthalpy of the diluting gas is also determined prior to injection into the stream.

Gases to be used as a diluent in this type of probe should be easily and accurately detectable quantitatively in the mixture with the gas whose properties are being determined. They should obviously be inert with respect to the incoming gas as well as to the materials of construction, to eliminate errors due to chemical reactions. Thermodynamic properties of the gas mixture which is drawn from the exit of this instrument must be known.

An enthalpy-sensor concept which has become more popular within recent years is the *transient or fast-response probe* [133]. The concept on which this sensor is based differs from the calorimetric types discussed up to this point in that it

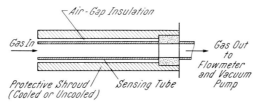

Fig. 66. Fast-Response Transient Enthalpy Probe

does not cool a sample of the hot gas to a point where conventional measurement techniques can be employed. Rather the basic principle of this probe operation is that of passing the incoming hot gas through a long, slender sampling tube for a short duration of time and obtaining a measure of the initial gas enthalpy from the temperature-time response of the tube.

The sensor consists of a sampling tube which is insulated from the outer protective shield by an air gap. The outer protective shield can be either cooled or uncooled. Fig. 66 shows the device in a schematic form.

When the probe is used, the hot gas to be sampled is permitted to flow through the sensor tube for a short time, increasing the temperature of the tube. If the heat transfer across the air gap is assumed to be negligible, the exit enthalpy at any instant is given by:

$$W_g H_{g_f} = W_g H_{g_o} - C_t M_t \frac{\delta}{\delta t} \left[\frac{1}{L} \int_0^L T_t(x)\, dx \right] \tag{34}$$

where:

C_t Specific heat of the tube material
M_t Mass of the tube
T_t Local temperature along the tube
L Tube length

In this fast-response transient instrument, the enthalpy of the exiting gas is changing with time and would be most difficult to determine with any degree of accuracy. With a properly designed sensor tube however, the final gas enthalpy is very low relative to the initial incoming gas enthalpy ($H_{g_f} \ll H_{g_o}$). It is the usual practice, when using this relationship, to neglect the term containing the final gas enthalpy (H_{g_f}). The relationship can then be rewritten and simplified to the following working equation:

$$H_{g_o} = \left(\frac{C_t M_t}{W_g} \right) \frac{d(T_t)_{\text{ave.}}}{dt} \tag{35}$$

where $(T_t)_{\text{ave.}}$ is the average temperature along the length of the tube. Equation 35 is an approximate relationship which holds over that portion of the heating of the tube in which the temperature rise is linear with time. In reality, the exit gas enthalpy can be considered as the reference or zero-enthalpy level when using this relationship.

There are two different methods in which this transient type of enthalpy probe can be used. One involves the rapid insertion and removal of the probe, while the plasma is operating under steady-state conditions. The other method fixes the probe in place while the gas flow is turned on and off again very rapidly. Although each of these techniques has some very obvious disadvantages, the fixed location method is reportedly the more practical of the two [135]. Either cooled or uncooled probes can be used in the fast-insertion mode of operating but only a cooled model can be used for the fixed-position method.

Despite the obvious difficulties of making accurate transient measurements of the gas-sample flow rate and the tube temperature rise, as well as minimizing the heat transfer across the air gap, users of this probe have obtained data having a standard deviation of 10% for enthalpies up to 2000 Btu per lb. The probe also demonstrated linear response to enthalpy up to about 3000 Btu per lb as indicated by calibration against other, more standard enthalpy probes.

Another interesting sensor design is the *sonic-orifice pneumatic probe*. The pneumatic probe has been defined [136] as a "thermodynamic temperature-measuring device which depends on the application of the Continuity Equation of a continuously flowing sample of the gas with unknown properties".

A simple application of the pneumatic-type probe is the double sonic-orifice shown schematically in Fig. 67. This sensor depends on an application of the steady-state Continuity Equation to a gas expanding through two sonic orifices in series. The concept is similar to that which applies to the sonic-flow method of determining the total enthalpy of a gas in a reservoir by expanding it through a nozzle. This technique is described in some detail in Chapter 6.

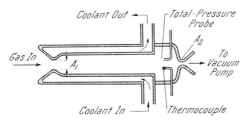

Fig. 67. Double-Sonic Flow Enthalpy Sensor

The analysis assumes that the gas under consideration is ideal in nature. This perfect gas of unknown stagnation temperature flows isentropically through the first nozzle. The flow is choked so that the gas velocity at the throat of the nozzle is sonic. The expanded gas enters a cooling chamber in which its temperature is reduced to a level which is measurable by standard thermoelectric or thermometric techniques. It is not necessary to know the amount of energy removed by the coolant, as it is with the calorimetric probes. The gas then flows isentropically through the second choked nozzle, and is exhausted through a vacuum system. Steady-state measurements of the initial total pressure in the unknown gas stream and the total pressure and temperature entering the second sonic orifice are made, and the two orifice areas, A_1 and A_2, must be known. In reality the effective flow cross-sectional areas, $C_1 A_1$ and $C_2 A_2$, are required, where the C's represent discharge coefficients which must be determined by calibration under simulated operating conditions. Then, by applying the Equation of Continuity to each choked nozzle, the mass flow through each can be written.

$$W_g = C_1 A_1 \frac{P_{g_o}}{\sqrt{T_{g_o}}} \sqrt{\frac{\gamma_o}{R_o} \left(\frac{2}{\gamma_o + 1}\right)^{(\gamma_o + 1)/(\gamma_o - 1)}}$$

$$= C_2 A_2 \frac{P_{g_f}}{\sqrt{T_{g_f}}} \sqrt{\frac{\gamma_f}{R_f} \left(\frac{2}{\gamma_f + 1}\right)^{(\gamma_f + 1)/(\gamma_f - 1)}}$$

or when combined

$$T_{g_o} = T_{g_f} \left(\frac{P_{g_o}}{P_{g_f}}\right)^2 \left(\frac{C_1 A_1}{C_2 A_2}\right)^2 \left(\frac{R_f \gamma_o}{R_o \gamma_f}\right) \frac{\left(\dfrac{2}{\gamma_o + 1}\right)^{(\gamma_o + 1)/(\gamma_o - 1)}}{\left(\dfrac{2}{\gamma_f + 1}\right)^{(\gamma_f + 1)/(\gamma_f - 1)}} \tag{36}$$

The subscripts o and f refer to initial conditions and conditions approaching the final nozzle respectively, and 1 and 2 refer to the nozzles.

In order to use this relationship to determine the stagnation temperature of the incoming gas the following conditions must exist:

(1) Perfect-gas, isentropic flow must prevail through each nozzle.
(2) The specific heat ratio (γ) and gas constant (R) of the unknown gas can be determined. It is assumed that these values for the cooled gas are given directly by its temperature, because the gas composition is either known or can be readily measured downstream.
(3) The orifice coefficients C_1 and C_2 must be known under operating conditions.
(4) The orifice areas A_1 and A_2 do not change between the time they are measured and the time of operation.

A major limitation of this technique is that although the cooling-chamber diameter is usually made sufficiently large so that the Mach Number in this region is far below unity (thus total and static values of pressure and temperature are approximately equal) the method as given above can only be used to determine the stagnation temperature of the unknown gas. If it is desired to know other equilibrium thermodynamic properties which depend on the free-stream temperature (e.g., enthalpy or density), it is necessary that either the unknown gas be flowing at low subsonic speeds (so that free-stream and total temperatures are nearly equal) or that the incoming Mach Number be measured separately. Also, if the unknown gas is partially dissociated or ionized, the assumption must be made that the degree of dissociation or ionization remains unchanged during expansion through the first nozzle and reaches equilibrium in the low-velocity cooling chamber between the two nozzles.

It is generally conceded that the double sonic-orifice probe is not well suited to measurements in reacting, nonequilibrium, partially dissociated, or partially ionized gases, or in environments which would produce changes in the orifice area of the first nozzle or its coefficient. This precludes its use with all plasma devices, hyperthermal wind tunnels, chemical combustion zones, and gases which form condensable solids. Even for reasonably high-temperature measurements in essentially perfect, noncondensing gases, the change in effective area of the first orifice is sufficiently large (up to 5% at 2500° F) to require prior calibration at or near operating temperatures.

In an effort to overcome some of these limitations, the triple sonic-orifice technique has been developed for use as an enthalpy-measuring device [136]. This probe is actually a combination of the pneumatic sensor and the diluent type discussed earlier. In operation the unknown gas enters through the first orifice and is cooled to relatively low temperatures by thorough mixing in an adiabatic chamber with a well-defined, cool diluting gas entering this chamber through a second choked orifice. This mixture then passes through the third orifice. Flow in all three orifices is sonic. Stagnation pressures and temperatures of the diluent gas and the cooled mixture are measured by conventional methods, as is the composition of the cool mixture exiting from the probe. The effective areas of the second and third nozzles must be known. These are both at low temperatures, however, and therefore not subject to significant change under operating conditions. Neither the state of the unknown gas nor the effective orifice diameter of the first nozzle is required, thus eliminating essentially all the limitations of the double-orifice probe. The first choked orifice serves only to isolate the mixing chamber.

The total enthalpy of the incoming gas can be determined from these measurements by combining the Conservation of Energy Equation with the Continuity Equation. The resulting relationship is:

$$H_{g_o} = \frac{w_m H_m - w_c H_c}{w_{g_o}} \tag{37}$$

And since the diluent (coolant) and mixture are at relatively low temperature, the relationship can be rewritten:

$$H_{g_o} = \frac{w_m C_{p_m} T_m - w_c C_{p_c} T_c}{w_{g_o}} \tag{38}$$

This is the working relationship which can be used with the triple sonic-orifice probe.

The only significant limitations on this device are the requirements for thorough mixing and equilibration of the gas entering the third orifice, and adiabatic conditions in the mixing chamber between the two upstream orifices and the third, downstream orifice. A minor limitation, as in the case of the double-orifice probe, is the requirement that adequate pressure ratios be maintained across all orifices to ensure the existence of choked flow.

There are several techniques for determining the enthalpy of a high-temperature stream which depend entirely upon the measurement of the heat-transfer rate from the gas to the probe. One of these techniques utilizes the measurements of heat flux across a calorimeter surface and stagnation pressure. Application of the theoretical heat-transfer analysis permits computation of the gas enthalpy required to produce the measured heat flux. This calorimeter technique requires a sensor to measure the stagnation-point heat flux, a probe to determine the total pressure and knowledge of the heat-transfer theory which governs the physical configuration being utilized. Calorimeters for the measurement of heat flux and pressure probes are discussed in detail later in this chapter. The theory of heat transfer at the stagnation point in a reacting gas and its relationship to enthalpy is pertinent for discussion here. The major difficulty arises from the fact that the flow is often hypersonic and that the gas must be considered reacting. The reactions which are taking place are primarily dissociation (with diatomic gases) and ionization. The rate of recombination at the surface where measurements are being made, has a very pronounced effect on the rate of heat transfer, of course. The situation is further complicated by the catalytic nature of the measuring surface, which is the predominant factor regulating the rate of recombination. The magnitude of this catalytic effect and its dependence on the type of materials involved will also be discussed later in this chapter.

Theories of boundary-layer heat transfer at the stagnation point from a dissociated gas have been formulated by several investigators. The theories of LEES, FAY and RIDDELL, GOULARD, SCALA, SIBULKIN, and ROSNER are discussed at some length in the literature [136, 137, 138, 139, 140, 141, 142]. It has become general practice to use FAY and RIDDELL's catalytic-wall analysis to determine the enthalpy of a dissociated gas from the measured heat transfer rate. FAY and KEMP's comparatively simple binary diffusion theory [143] appears to describe the added complexity of ionization most accurately.

There has been considerable discussion regarding the effect of the state of the gas within the boundary layer on the heat transfer. LEES and ROSNER have stated that the state within the boundary layer is not important but that the heat transfer for given free-stream conditions depends only on the condition at the wall or measuring surface, regardless of whether that condition resulted from recombination within the boundary layer or on the surface itself. It is for this reason that the catalytic activity of the surface of a calorimeter is so important in determining heat transfer rates. It becomes more important as the conditions within the boundary layer are farther from thermal equilibrium.

Because of the excellent correlation shown by the simple binary theories for heat transfer in both dissociated and ionized gases, several simplified models have been postulated. ROSNER has developed one such relatively simple model which is essentially in agreement with the detailed numerical solution of FAY and RIDDELL for heat flux from an equilibrium dissociated gas [144]. This model apparently yields good estimates of the relationship between the gas enthalpy and the heat-transfer rate. The parameters which are required to make use of the relationship include: measured values of the temperature of the sensing surface of the calorimeter and the heat flux, and estimated values for the "frozen" Lewis Number, the Prandtl Number and the mass velocity. A knowledge of the catalytic nature of the calorimeter surface is used to provide a value for enthalpy at the wall while the Equation of State is used to provide a value for free-stream enthalpy. This technique therefore depends not only on the accurate measurement of the heat flux and wall temperature but also on the catalytic activity of the measuring surface and on the assumption of certain characteristics of the free stream.

Another technique which depends on the heat transfer to a sensor to deduce gas enthalpy is the *ablating probe*. This technique determines the ablation rate of a material which has been precalibrated to determine its ablation rate as a function of stagnation-point heat transfer. The heat transfer rate to the probe is given by:

$$q = \dot{m}Q_a \tag{39}$$

where $\dot{m}$ is the mass ablation rate and Q_a is the heat of ablation for the probe material. The heat of ablation of a material is defined as the amount of incident energy which can be dissipated by the ablation of a unit mass of a material. By using a probe material of known heat of ablation it is a simple matter to determine the heat transfer rate to the probe by measuring the mass ablation rate. If a rod of constant cross section is mounted with its major axis parallel to the direction of flow, and if the heat of ablation and density of the rod are known, then the measured longitudinal rate of ablation, together with the measured probe-surface temperature provides the information required for application of the stagnation-point heat transfer analysis.

Although this technique is by far the simplest and least expensive of all enthalpy probes, it suffers from a number of inherent errors. Most important, the heat of ablation itself is a function of the unknown gas enthalpy. The three-dimensional heat flux and ablation of the probe's lateral area must be either accounted for by calibration or minimized by the use of a cooled or sacrificial ablating shroud, into which the probe is fed at a rate equal to its longitudinal ablation rate. Although lateral ablation is thus eliminated, three-dimensional heat transfer to the shroud must be calibrated. Also, the finite area of the probe reduces the applicability of the simple stagnation-point heat transfer analysis, and this finite area, together with the probe surface regression rate, also reduces the usefulness of the probe in flow with steep temperature gradients. Despite these difficulties, however, the ablating probe is frequently utilized for quick and inexpensive approximations in high-temperature gas flows.

7.2. Calorimeters

Another diagnostic device which is used to define the thermal environment of the flowing test gas is the heat-flux gauge or calorimeter. Although the concepts and designs of the various calorimeters vary widely, they have one common function; all measure the heat transfer from the hot flowing stream to the sensor. The data which are gathered with a calorimeter can be used, with the aid of the proper analytical manipulations discussed in the preceeding section, to calculate the enthalpy of the hot gas. More generally, however, the output of a heat-flux sensor is used simply to determine the heating rate with little or no consideration being given to the values of the individual parameters which resulted in that particular heat-transfer rate or heat flux. Corrections to the experimentally measured values must usually be made before the heat flux to the test-model surface is known, but this parameter, specimen heat flux, is generally of prime interest in materials evaluations.

Heat flux has been determined by many different concepts and even more different individual instrument designs. One of the many calorimeter concepts which has been popular for use in materials-testing facilities is the *slug calori-*

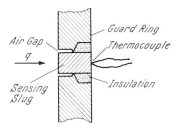

Fig. 68. Transient Slug Calorimeter for Measuring Heat Flux

meter, more properly called the *thermal-capacitance calorimeter*. A schematic sketch of this concept is shown in Fig. 68. The heat sensor or slug is constructed of a material of well-defined thermophysical properties. It is insulated from the body or guard ring in which it is supported. A thermocouple is attached to the back face of the slug and is used to register the temperature of that surface. The opposite (front) surface of the slug is exposed to the unknown environment and it is the heat transfer rate across this surface which is of interest. The insulation which minimizes the heat transfer out of or into the cylindrical surfaces of the slug is very important since the assumption of one-dimensional (axial) heat flow is made with this calorimeter.

To determine the heat flux with this instrument, the sensor is inserted in the flow field. As heat is transferred to the slug the temperature of any point in the slug increases. By measuring the rate of rise of a specific point (the plane of the rear surface) a value for the incident heat flux can be calculated. The time-temperature response of the back surface which is recorded is typified by a short "lag time" followed by smooth transition to a linear temperature increase which continues for a short time after the heat input to the slug has discontinued. It is this linear increase in temperature which is of interest and is used in the heat flux calculation. The energy is transferred from a high-temperature gas to a relatively cold surface. The temperature of this surface increases immediately, and theoretically the heat flux to the surface begins to decrease. Since there is such a large difference through which the heat is transferred, the temperature rise of the front surface during these short-time exposures is very small by comparison. The result is that there is a short time when the heat transfer rate to the slug is essentially constant. It is this value of heat flux which is measured.

An energy balance can be written to define the conditions during this time of linear response which states that the rate of energy transfer across the front surface (heat flux) is equal to the rate of energy transfer by conduction axially into the slug [145]. The time-dependent temperature response at any point in the slug is given by:

$$q = \frac{Q}{A} = \varrho\, C_p x\, \frac{\mathrm{d}T}{\mathrm{d}t} \tag{40}$$

where:

A Area of front surface
C_p Specific heat of slug material
t Time
T Temperature
q Heat flow rate/unit area (Heat flux)
Q Total heat flow rate
x Distance from heated surface
ϱ Density of a slug material

Since the thermocouple used to measure the temperature response of the calorimeter is placed on the back surface of the slug, the entire mass of the slug is heated. The working equation for the temperature-time response of the back surface can thus be written:

$$q = \frac{MC_p}{A}\,\frac{\varDelta T}{\varDelta t} \tag{41}$$

In this equation M is the mass of the cylindrical slug. The quantity $\varDelta T/\varDelta t$ is the slope of the recorded time-temperature response which is measured over the linear portion of the curve.

Since the validity of this technique is dependent on the one-dimensional (axial) transfer of heat through the slug, it is essential that the heat transfer across the side walls of the slug be extremely small. The physical separation of the slug from the body of the calorimeter by means of an air gap and/or a good thermal insulator is very important. The insulating air gap should be no more than approximately two mils thick. If this gap becomes too large, pressure variations which may exist across the face of the calorimeter can produce side heating by the flow of hot gases into and out of the gap.

The only important parameter to be measured when using this type of instrument is the back-surface temperature. The method for measuring this temperature must be sufficiently sensitive and reliable to insure accurate temperature-rise data for the back-face thermocouple. Procedures similar to those given in the ASTM Standards should be adhered to in the calibration and preparation of the thermocouple. Attachment of the thermocouple should be such that the true back-side temperatures are obtained. Although no accepted procedures are available, methods such as resistance welding (small spot) and peening have proved preferable. The temperature measurement should be recorded continuously, using a recorder whose response time is sufficient to provide the accuracy required.

There are several empirical or semi-empirical relationships [145] which are useful when working with a slug calorimeter. A nomogram is available to aid in selection of a slug material, exposure time, front-surface temperature rise for a given heat flux, etc. The following relation is given for calculating initial response (lag) time for a given slug:

$$t_R = \frac{L^2}{2k/\varrho C_p} \tag{42}$$

where k is the thermal conductivity of the slug and L is the length. The optimum length of the slug can be obtained from:

$$L_{\text{opt}} = \frac{3}{5} \left[\frac{kT_{\text{front}}}{q} \right] \tag{43}$$

The maximum amount of linear response time is:

$$t_{\text{max}} = 0.48 \, \varrho k C_p \left[\frac{T_{\text{max}} - T_o}{q} \right]^2 \tag{44}$$

The calorimeter and its holder or surrounding body should, whenever possible, be of a size and shape identical to that of the model which is to be tested. Only in this manner can the measured heat flux to the calorimeter be related to that experienced by the test model. With appropriate modifications this basic principle of a thermal-capacitance calorimeter can be used to measure the heat flux on various parts of larger test models. This is accomplished by instrumenting the model with calorimeter slugs, thereby permitting the measurement of heat flux during an actual test.

Regardless of the source of thermal energy to the calorimeter (radiative, convective, or a combination thereof) the heat flux measured is an average over the calorimeter surface. If a significant percentage of the total thermal energy is radiative, consideration should be given to the emissivity of the slug surface. Also, if non-uniformities exist in the input energy, the calorimeter tends to average these variations; therefore, the size of the sensing element (slug) should be limited to small diameters in order to measure local heat-flux values. Where large samples are to be tested, it is recommended that a number of calorimeters be incorporated in the body of the test specimen such that a heat-flux distribution across the heated surface can be determined. In this manner more representative heat-flux values can be defined for the test specimen, thus permitting more meaningful interpretation of the test [145].

The slug calorimeter is inexpensive and simple to fabricate and has therefore been quite popular. The major disadvantages are those it shares with all other transient-type diagnostic probes. Since it is non-steady state in operation, the probe must be withdrawn from the effluent and permitted to cool, after each reading, before it can be used again. This can be a nuisance, particularly when an attempt is being made to adjust the operating conditions of a plasma jet to produce a particular heat-flux level. The necessity for removing the instrument, waiting and then reinserting it after each minor adjustement can be inconvenient. Another disadvantage of this particular type of calorimeter is the relatively short life of the slug and the tendency of the instrument to lose calibration due to changes in the contact of the thermocouple with the slug surface.

Another approach to the one-dimensional heat transfer concept of determining the heat flux is illustrated by the *thin-skin calorimeter*. This instrument makes use of the same time-dependent temperature-response relationship as was developed for the slug calorimeter (Equation 40). The method of attaining the condition of one-dimensional heat flow differs from that used in the slug technique. In the thin-skin concept the rate of temperature rise is measured on the back surface of a very thin sheet of metal which is the sensing area. By using a thin metal sheet of relatively large area, the heat transfer is assumed to be through the thickness of the sheet. Thermocouples are attached to the back surface of the sensing area by spot, electron beam or laser welding. The two thermocouple elements (wires) are located on an isothermal line to assure accurate temperature measurements. Normally, several thermocouples are used with these instruments.

These calorimeters are easily constructed in a configuration and size similar to the model to be tested in the environment. By recording the rate of temperature rise of many properly placed thermocouples simultaneously, the distribution of incident heat flux can be determined with a single exposure. This generally requires the use of a relatively low thermal conductivity metal for the sensing area, particularly if the heat flux gradient across the calorimeter surface is large. Because of their thin materials of construction these probes cannot be operated in high-pressure atmospheres (or for that matter at moderate pressures, if the temperature becomes too high) since severe distortion of the sensing area will occur.

A direct heat-transfer measuring instrument which has also been popular over the past decade is the *asymptotic calorimeter*. This instrument, which is also known as a circular foil or Gardon gage calorimeter, makes use of the temperature difference which exists over a finite thickness of any material through which heat is flowing. By designing the instrument so as to assure heat flow through a desired path, the temperature difference which exists is directly proportional to the heat flux. A schematic drawing of a typical asymptotic calorimeter is shown in Fig. 69.

The unit consists of a thin-gauge metallic foil which is suspended over a cavity in a heat sink. This thin metal portion is the sensor. It is bonded or joined to the heat sink around its periphery to provide good thermal contact. With this configuration the incident heat is directed radially outward to the heat sink. The heat sink, which usually acts as the main body of the probe, may be either liquid-cooled or uncooled. The steady-state heat flow results from a temperature gradient which exists between the center of the sensing disc and any point located out toward the periphery. The magnitude of this temperature differential is

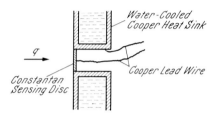

Fig. 69. Asymptotic Heat-Transfer Sensor

directly proportional to the incident heat flux to the foil. By constructing the sensing foil from a thermoelement metal alloy, such as constantan, and the heat sink of another thermoelement, such as copper, the periphery where the two are joined (assuming there is good electrical contact) becomes a thermocouple junction. A copper wire attached to the center of the foil sensing disc provides another thermocouple junction. The combination of these metals in this manner results in a differential thermocouple circuit which has the center point of the disc as the hot junction and the periphery as the cold junction. The millivolt output from this circuit gives a precise measure of the temperature differential between the center and the outside edge of the foil without actually measuring either temperature. It can be shown that this temperature differential is directly proportional to the heat input over the entire area of the sensing disc.

By performing an energy balance over a washer-shaped section, Δr wide, through the thickness of the foil, L, the following equation can be written [146]:

$$[2\pi rLq_{r_i}] + [2\pi r\Delta rq] = [2\pi(r + \Delta r)Lq_{r_o}] \tag{45}$$

This equation states that the energy conducted radially into the section (q_{r_i}) plus the incident axial heat flux to this section (q) equals the energy conducted radially out of the section (q_{r_o}). Rearranging this expression and taking the limit as the width of the washer section (Δr) approaches zero, results in the following first-order differential equation for the energy flux:

$$\frac{d(rq_r)}{dr} = \frac{rq}{L} \tag{46}$$

Integrating and applying the boundary condition that q_r cannot be infinite at $r=0$ yields:

$$q_r = \frac{qr}{2L} \qquad (47)$$

To obtain a temperature distribution, Fourier's Law is substituted for q_r to obtain:

$$-k\frac{dT}{dr} = \frac{qr}{2L} \qquad (48)$$

This is also a first-order differential equation if the thermal conductivity k is constant. Integrating this equation between the limits $r=0$ and $r=R$ results in the following expression for temperature distribution across the surface of the disc:

$$(T_o - T_R) = \frac{q}{4Lk}(R)^2 \qquad (49)$$

or in terms of the incident heat flux:

$$q = \frac{4Lk}{R^2}(T_o - T_R) \qquad (50)$$

This equation is the working expression which defines the heat flux to an asymptotic calorimeter as a linear function of temperature differential from the center to the periphery of the sensing disc. The above derivation assumes that the incident heat flux is constant over the entire area of the disc and that the thermal conductivity of the metal from which the disc is constructed does not vary with temperature.

An advantage of the asymptotic-type calorimeter is that, since the output is directly proportional to the heat flux, data reduction is rapid, simple and inexpensive. They also display a fast response time for a steady-state device, on the order of one-quarter second or less [147]; they can be made very small in size, are reasonably rugged and are insensitive to variations in ambient temperature. Because of their simplicity and lack of auxillary equipment, these probes have found many uses outside of the laboratory or test facility. Individual models are limited to relatively small variations in the heat-flux levels over which they can operate. They must be designed in such a manner as to have adequate cooling in conjunction with proper sizing of components to prevent damage to the thin foil disc during use in high-flux environments, yet they must produce sufficient temperature gradients in the less severe environments to permit accurate measurements.

The most widely used steady-state instrument for determining heat flux in an arc plasma effluent is the *water-cooled calorimeter*. A typical water-cooled calorimeter is shown in Fig. 70. In this instrument water flows down the center tube toward the sensing face. It passes across the back of this sensing area and returns down the outside channel. Thermocouples are placed in the water stream to determine the temperature rise which results from passing over the sensing area.

A steady-state heat balance reveals that the heat input to the sensing area per unit time (heat flux) is equal to the heat which is carried away by the cooling water. The value of the heat flux is found by determining the amount of energy which is absorbed by the water as it cools the sensing area. The equation used to compute heat flux is:

$$q = \frac{(mC_p \varDelta T)_{\text{coolant}}}{A} \tag{51}$$

As is the case with many other calorimeter designs as well as most enthalpy probes, the effects of heat transfer from sources other than the gas sample of interest must be minimized. In the model shown in Fig. 70 the entire sensing unit is protected by a water-cooled main body or by a guard ring. The sensing tube is isolated from this cool guard ring by a small (one to two mil) air gap.

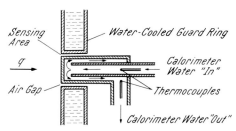

Fig. 70. Water-Cooled Calorimeter

As stated earlier, maintenance of this air gap at approximately these dimensions is essential for reliable probe performance. Another design, which is intended to lessen the effects of extraneous heat transfer, places the thermocouples very near the sensing surface. A disadvantage of this technique is that the temperature determined at this close proximity to the point where it is heated may not be the "mixing bucket" or bulk temperature which is required for use of Equation 51. If the water is permitted to flow for some short distance before the temperature is measured, there will be a much greater chance of the "averaging" tendencies to occur.

A particular advantage of the water-cooled calorimeter is that a single unit can be used to measure a wide range of heat-flux values. Because the flow rate of water through the instrument can be varied, a significant temperature rise in the water can be maintained. This results in an instrument with which accurate measurements can be made over a heat-flux range of several orders of magnitude. Practical considerations, such as minimum thickness of the metal in the sensing area, generally limit the water-cooled calorimeter to heat-flux levels of less than 10,000 Btu per ft^2 sec.

Water-cooled calorimeters are extremely slow to respond to changes in the incident heat flux. Since the parameters are measured over a finite period of time, very steady heat-transfer conditions must exist to obtain accurate measurements. Also, because they tend to be cumbersome and require a considerable amount of auxiliary equipment to operate, these units are seldom used outside the laboratory test facility. They are generally used only as probes and are rarely incorporated into a test model.

One of the more recent developments in steady-state heat-flux gauges is the *multiple-wafer calorimeter* [148]. Like the asymptotic calorimeter, the wafer heat-flux gauge makes use of the dependence of the heat transfer on the temperature gradient through a section of some known material. This method makes use of Fourier's heat-conduction equation in the simplest form, i.e., that of one-dimensional heat flow.

The typical wafer-calorimeter configuration is shown in Fig. 71. It consists of three thin, flat metal wafers or discs which are diffusion-bonded together. As with the asymptotic type of instrument, the use of thermocouple materials such

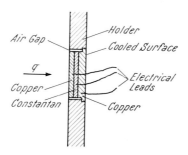

Fig. 71. Multiple-Wafer Calorimeter

as copper and constantan, for these discs results in the disc interfaces forming thermocouple junctions. When the front surface is exposed to the high-temperature gas stream while the back surface is maintained at a lower temperature (usually by water cooling) a temperature gradient across this sandwich of wafers is established. The thickness of the center wafer is much smaller than its diameter so as to assure that the heat flow is one-dimensional. Since the wafers are constructed from thermocouple materials and form junctions at the wafer interfaces, the temperature at each face of the center wafer can be accurately determined. Knowing these temperatures, the mean thermal conductivity of the center wafer can be obtained from thermophysical property data for the particular wafer material. The thickness of the center wafer is very accurately measured at the time the calorimeter is assembled. The measured temperature drop, ΔT, the known wafer thickness, L, and the mean thermal conductivity, k, may then be used in the simple integrated form of the one-dimensional Fourier Equation to calculate the heat flux, q:

$$q = \frac{k}{L}(\Delta T) \tag{52}$$

The calorimeter is intended to simulate as nearly as possible the condition of this one-dimensional Fourier heat-conduction equation, by attempting to match the "effective" thermal conductance of the calorimeter with the thermal conductivity of the base material or cooled wall in which it is mounted.

The wafer calorimeter is a steady-state device intended primarily for applications where it will be installed in a cooled wall. Although a coolant is required for maintaining the low temperatures of the back surface, its flow rate and temperature rise need not be measured to obtain the heat flux, since the only operating measurements that enter into the calculation are the temperatures at the wafer interfaces. Also, the relative thicknesses of the wafers can often be varied to match exactly the conductance of the base material in which the calorimeter is mounted. Local distortions in heat flux are minimized in this manner, and the accuracy of the measurement improved.

The perimeter of the calorimeter is insulated to avoid lateral heat conduction to the base material, which would void the one-dimensional heat-flux assumption. The front-surface wafer and center wafer should also be electrically insulated from the adjacent base material. This is generally accomplished with an air gap of two to three mils surrounding the calorimeter. The electrical leads should be as small in diameter as possible to minimize heat loss from the wafers along these routes.

The accuracy of the heat-flux measurements made with the multiple wafer calorimeters is dependent upon the design and fabrication techniques employed, as well as on the accuracy of the temperature measurements. Certified high-quality thermocouple-grade materials should be used to construct the wafers. The center wafer should be very accurately measured (to the nearest 0.00005 inch, if possible) with an optical comparator, after the wafers are bonded. A properly designed water-cooled model of this type of calorimeter is capable of measuring heat-flux values up to 5000 Btu per ft^2 sec.

The multiple-wafer heat-flux gauge is a steady-state device and, as such, exhibits the advantages characteristic of this mode of operation. It also has some of the same disadvantages as the water-cooled calorimeter, primarily that of slow response time and the requirement of very steady heat inputs.

One of the transient-heat-flux meters which has come into increased use in recent years is the *null-point calorimeter*. Additional interest has been shown in this concept because of its adaptability to use in high-heat-flux environments such as those encountered in high-pressure arc tunnels. The null-point concept is based on the principle that there is a unique thermocouple-hole depth, below the heated surface of a material, at which the measured temperature is equal to the material-surface temperature, after a short time interval [149]. The null-point calorimeter is constructed so that the temperature history measured by a thermo-couple is identical to the temperature history of the undisturbed surface of a semi-infinite solid of the same material exposed to the identical incident heat flux. It has been shown [150] that under the proper conditions this can be achieved at the bottom of a hole drilled in a semi-infinite body to such a depth that the bottom of the hole is a distance equal to one hole radius from the heated surface. The heat flux is calculated from the time history of the measured temperature.

The details of a typical null-point calorimeter are shown schematically in Fig. 72. The cylindrical body or slug shown is installed in a model of nearly any desired shape and dimensions. Although the construction of these calorimeters appears to be very simple, they are nearly always very small and therefore difficult to machine and assemble. The hole, for example, must be normal to the heated surface and the bottom must be flat for the null-point concept to be valid. The thermocouple is soldered or brazed to the bottom of the hole. The slug is constructed of a high-thermal-conductivity metal such as copper or silver. A conduction break or air gap is usually machined into the slug body, but the front face must fit tightly into the body (press fit) to prevent gas circulation in the gap at the high pressures associated with the use of this instrument. The diameter of the slug is large compared to the diameter of the hole so that the slug serves as a semi-infinite body in the radial direction. If the slug diameter is not considerably greater than that of the hole, the null-point requirements will be violated and a significant error will result.

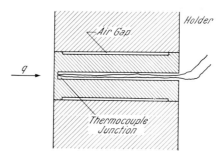

Fig. 72. Null-Point Calorimeter

The values for the heat flux are calculated with the aid of one of several computer programs which have been written to analyze the indepth heat conduction in a solid body with non-uniform surface heating. Some programs have been specifically written to apply to the null-point calorimeter while others are adaptations of programs intended primarily for calculating heat transfer and ablation performances of such things as nose tips and rocket nozzles. Most make use of the finite-difference technique in some manner.

One such program was used to determine the errors which are introduced by various deviations from null-point requirements [149]. Among these was the effect of hole-to-slug size ratio. This effect was found to be relatively small for ratios of 0.2 or less. Similar investigations revealed that a small quantity of solder at the bottom of the thermocouple hole will not significantly affect the response for a null-point calorimeter. Potentially more important effects are due to the alloying of the solder (or braze) either with the slug material or the thermocouple wires. Alloying with the slug material can effect the thermal response of the instrument while alloying with the thermocouple wires can alter the thermoelectric character of the junction. The heat conduction away from the junction along the thermocouple wires has a negligible effect on the recorded temperature response. This is dependent upon the materials of construction and the heat flux values, but for materials such as copper, silver and constantan and for high heat-flux levels, all of which are normal for these instruments, the effect is very small.

The advantage of this type of calorimeter is the small size attainable, which allows point-by-point mapping of the test section and measurement of very high heating rates. They are used either in the fixed position or in the sweeping mode of operation. The use of the sweeping mode offers the advantage of repeated use of an instrument and the ability to map the heat-flux profile during a single exposure. Primary disadvantages, in addition to those associated with all transient calorimeters, are the relative difficulty of analyzing the data and computing heat-flux values.

7.3. Pressure Probes

The pressure of the environment to be used in evaluating the performance of a material is another parameter which is nearly always of interest. Both static (local) pressure and stagnation (impact, total) pressure are often useful when analyzing the behavior of a model during test. Unlike most of the other test parameters which are measured in connection with a materials test, pressure determinations require very little specialized equipment or techniques. Accurate measurements can be obtained at relatively high gas-stream temperatures with existing techniques. As gas temperatures are increased to the very high values associated with an arc plasma facility, the basic probe designs are complicated slightly by the necessity of providing protective cooling for the instrument. Cooled pitot probes are widely used for determining both static and stagnation pressure in test facilities.

These *impact-type probes* are subject to error from at least two sources when used in a low-density, high-temperature flow field [151]. Viscosity effects on the measured impact pressure can result from the low Reynolds Numbers which can exist in a low-density environment. Also, a phenomenon known as thermal creep can effect the readings. Thermal creep is due to the temperature gradient along the pressure line connecting the orifice and the pressure transducer. This can usually be made negligible by the proper sizing of tubes for the specific environment. The viscosity effects can be lessened by increasing the size of the probe but unfortunately this increases the thermal-creep effects. In other than low-density plasma streams, however, these probes are relatively free from complications.

An instrument which has been devised for use in measuring short duration phenomena is the *piezoelectric probe* [152]. This probe can be used to determine very small pressures (as low as 0.01 atmospheres) and fluctuations on the order of microsecond durations. Since this probe makes use of a piezoelectric crystal with characteristically small signal outputs, difficulties are often encountered with signal distortion. *Diaphragm-type pressure indicators* have also been devised for measuring dynamic pressures [153]. These are rather easily affected by changes in temperature, even though they are water-cooled, and their usefulness in plasma streams is limited.

7.4. Velocity Measurement

The velocity of the flowing gas can also be determined from measurement of stagnation and static pressure. The pitot tubes, which are the most widely-used probes for determining stream velocity, are designed in the usual manner with the exception of the necessary water-cooling provisions. These probes can easily be miniaturized and are capable of measuring the velocity at precise locations. They can therefore be used to map a cross section of the stream, resulting in velocity profiles rather than average flow velocities.

Velocity measurements can also be made with electrical probes such as the Langmuir probe. The *Langmuir probe* is essentially a charged-particle collector. When this device is placed in a plasma and maintained at varying potentials with respect to the plasma, the resulting current-voltage relationship can be plotted. These data are generally used to calculate ion and electron temperatures and densities. However, by mounting two Langmuir probes in a position such that one is downstream from the other, a measure of average stream velocity can be obtained. When such a double probe is used, the outputs are usually recorded on an oscillograph. The velocity is calculated from the time between identical signals from the two probes being recorded on the oscillograms. Hot-wire anemometers and other related thermal-type velocity meters have also been used to measure velocity. They are limited to very few applications in plasma-heated test streams, however, because of the high temperatures encountered.

Another type of velocity probe which has been used [154] makes use of the *Faraday effect*. In this technique a constant magnetic field is set up perpendicular to the direction of flow, and an emf is induced in the plasma stream. The emf is detected by adjustable probes or pick-ups and is transmitted through a pulse transformer to an oscilloscope. The velocity is calculated from the following relationship, which was obtained from Faraday's Law:

$$E = BVd \tag{53}$$

where:

E Potential between probes
B Magnetic field strength
V Velocity of charged particles
d Distance between probes

Since the average velocity at any location in the stream can be measured with this procedure, velocity profiles can be obtained. It is applicable to any ionized gas stream, but fails when the gas stream is not ionized. Errors are introduced by contamination from ablation of the pick-up electrodes and from impurities associated with slower particles.

There are several techniques utilizing *photographic techniques* to obtain direct measurements of stream velocity. One of these involves the use of a rotating-drum camera. The camera drum is rotated about an axis parallel to the direction of the flow. The film is thus swept perpendicularly by the stream resulting in oblique streaks on the developed film. Measurement of the angle of inclination of these streaks, with knowledge of the rotational speed of the drum, permits computation of the average velocity of the stream. Micron-sized particles can be added to the stream to act as tracers [155]. This permits the camera to be focused on narrow regions of the jet to obtain velocity profiles. This raises the question of motion of the particles in the free stream. Even with very low-density particles, the measurements must not be made until the residence time has been sufficient to assure equal velocity between particles and stream. If the velocity is not particularly high, the effects of gravity acting on the particles (in a horizontally flowing stream) can introduce a sizeable error by changing the angle of the streaks on the rotating film.

Another technique simply measures the velocity of luminous "blobs" of plasma [154]. This is accomplished by placing two photomultiplier tubes several inches from each other in the direction of plasma flow. The output of each sensor is transmitted into a separate readout channel consisting of a pulse-shaping amplifier and a "read" amplifier. As the flow of plasma gas passes in front of each sensor, luminous fluctuations in the flow are seen as variations in amplitude at the output. The outputs of the two sensors are identical to each other with the exception of a time delay. A precise measure of the time required for the flow to travel the distance between the two sensors can therefore be obtained and the average velocity calculated.

Although time intervals down to two microseconds can be accurately measured, the method encounters serious difficulties if the stream flow is not essentially uniform. If luminous particles such as result from electrode contamination are present, erroneous velocity readings can be obtained.

The spectroscope has also been used to determine stream velocity [156]. In observing the frequencies of spectral lines which are emitted, the *Doppler shift* which occurs may be used to measure velocity. The frequencies are observed at two different angles as close as possible to the upstream and downstream directions of flow. The relationship which can be used to determine velocity is:

$$\frac{\delta\lambda}{\lambda} = \frac{V}{C}(\cos\theta_1 - \cos\theta_2) \tag{54}$$

where:

θ_1 and θ_2 Angles of observation
$\delta\lambda$ Doppler shift
λ Frequency
V Velocity of stream
C Velocity of light

For light in the visible range and for gas velocities of about 6700 ft per sec, the shift of frequency is thus of the order of 0.1 Å and is directly proportional to the velocity. The lower sensitivity limit of this technique is estimated at 300 ft per sec.

7.5. Density Determinations

Another gas property which is of interest in materials testing is the density of the flowing stream. This parameter can be easily determined by using several of the probes previously discussed. The pressure-impact probe provides a particularly simple method if the velocity of the gas stream is known. The accuracy of the measurement of density by this method is actually limited by the accuracy with which the velocity has been determined. The same is true for density determined from mass-flow measurements. These measurements, which can be made with most of the enthalpy probes discussed earlier, are limited by the accuracy of the velocity measurements. The mass-flow technique has an advantage over the impact-probe in that it depends on the first power of the velocity term while the latter varies as the square of the velocity.

The change in the *index of refraction* of a gas with density also can be made the basis of a density measurement [154]. Interferometers are used to determine the difference between the optical-path length of a light ray passing through a given thickness of gas and a standard optical path length. If the variation of the index of refraction with density is known, the average density over the path length is immediately available from the interferometer measurement.

The major disadvantages associated with interferometry are the relatively high densities which are necessary for easy application and the difficulties with boundary layers on windows and other optical components. A further disadvantage is the relative lack of knowledge of the specific refractivity of high-temperature gases, and the necessity for knowing the concentration of species in the gases before the variation of index of refraction with density can be determined.

Schlieren and *shadowgraph* techniques may also be applied. However, schlieren techniques measure the first derivative of density and shadowgraphs indicate the second derivative of the density, whereas interferometer measurements give the density directly. Optical techniques allow density measurements to be made without introducing disturbances into the gas stream. If photographic techniques can be employed, optical techniques also allow the density to be measured over a large region of gaseous flow simultaneously and with short exposures so that transient phenomena may be recorded.

Other techniques which have been either discussed or used with limited success in determining density include spectroscopy, electron and alpha-particle attenuation, X-ray and spectral absorption, X-ray scattering and speed-of-sound determinations.

Spectroscopic techniques, although feasible in principle, are very difficult to apply to the determination of mass density. For one thing, local thermodynamic equilibrium is absolutely essential before the technique can be applied. *Electron-beam techniques* have been successfully used in determining the density of very rarefied gas flows. One method measures the beam attenuation on traversing the gas stream [157]. The apparatus used consists of an electron gun which produces a collimated beam of electrons, and a detector which records only those electrons not scattered while passing through the gas. The detector must have a narrow angular aperture in order to accept only those electrons which have survived the entire passage through the gas stream without deflection. One difficulty with this technique is that extraneous electric and magnetic fields easily alter the accurary of the measurement by substantial deflection of the beam so as to make the application of this technique to plasma-jet streams difficult. Similar *alpha-particle attenuation techniques* are feasible only when the density is fairly homogeneous along the path of the beam [158]. For situations where such a probe is applicable, high accuracies in density measurement are possible almost irrespective of the thermodynamic state of the gas.

The *absorption techniques* (X-ray and spectral) are useful only if additional knowledge about the gas is available [158]. The effects of such things as gas temperature and local molecular composition must be considered. Scattering of X-rays on the other hand, has been used to determine the electron density in the gas rather than the chemical state of the gas. The energy of these X-rays is sufficiently high that the scattering is from both free and bound electrons. The gas density is determined from this measurement of total electron density.

7.6. Gas Temperature

The question of gas temperature often arises when discussing high-temperature evaluation techniques using the arc plasma. The energy present in a plasma gas is distributed among various possible modes. These include translational, rotational or vibrational energy, energy of excitation, ionization or dissociation, and energy of radiation between the particles [154]. The concept of temperature for such a system may be introduced only when the energy content is distributed over the possible modes of the system so that equilibrium exists among them. Equilibrium for a gas at a given temperature implies that the number of all processes tending to populate any given state are equal to the numbers of all processes tending to depopulate those same states. For a gas in a real physical situation, such equilibrium cannot exist since there will be inevitable energy losses which must be balanced by energy supplied to the system.

The interpretation of the temperature measured by an instrument or technique must be made by taking into account the state of excitation of the test gas. Temperatures are often determined from the measured values of enthalpy, but these imply a complete knowledge of the state of the gas and the degree of equilibrium. Direct temperature readings have been measured by thermocouples and various resistance-thermometry techniques but these are limited by the upper use temperature of the materials from which they are constructed. Major errors can be introduced when using any technique which requires a sensor to achieve the temperature of the stream and then determines the temperature of this sensor. When thermal equilibrium is reached there will be losses from the sensor by radiation, stem or holder conduction, etc. which will not permit it to achieve the temperature of the gas with which it is being heated. The electrical charge which can build up on a probe placed in a highly ionized stream can also result in erroneous readings, particularly when thermocouples are involved.

The *Langmuir probe*, which was discussed earlier as a velocity-measuring device, has long been one of the standard techniques for determining electron temperatures and ion densities in a plasma [154]. The double probe [151] is generally used to avoid drawing excessive electron currents from the plasma. By measuring the gas pressure and ion density, the temperature of the gas may be inferred at equilibrium from the equilibrium thermodynamic composition of the gas corresponding to that ion density and pressure. The comparison of electron temperature with heavy-particle temperature in the gas gives a measure of the degree of equilibrium of the electrons and heavier particles in the gas.

Considerable effort has been expended on the application of spectroscopy to determine the temperature of hot gases. In general, spectroscopic methods of temperature measurement utilize techniques based on determining one of the following quantities [154]:

(1) Absolute intensities of spectral lines.
(2) Relative intensities or intensity ratios of spectral lines, or distribution of intensities of lines in a rotation or vibration band.
(3) Intensities of continuum radiation.
(4) Spectral line broadening or frequency shift.

For all of these techniques, the assumption is made that the medium is optically thin. Corrections can usually be made, however, to permit the use of data obtained from optical media of intermediate thickness.

Optical techniques, on the other hand, depend upon the fact that emission from an optically thick layer of a gas obeys Planck's law for black-body radiation. Thus, by intensity measurements from a gas in a frequency region where the gas is known to be optically thick, the temperature of the gas may be determined by the application of Planck's law. Optical techniques which have been used to determine temperature in this manner include the determination of the wavelength of maximum radiant-energy emission (Wien displacement law), two-color and total-intensity radiation pyrometry.

There are a number of methods which have been devised to measure high gas-stream temperature but have not been widely used. These include the velocity-of-sound method, microwave-attenuation measurements, thermal-neutron scattering, and measurement of microwave radiation from the plasma stream.

7.7. Chemical Composition

The chemical composition of the test gas, in particular the state of the atoms and molecules, is of interest in many materials research projects. Even with a plasma effluent which is generated from a singular gas species the gas may be present in several different states. Included may be gas molecules, some of which have dissociated to atoms (in the case of a diatomic gas) which may in turn be ionized, releasing free electrons. The quantitative measurement of the localized concentrations of species in any flowing, reacting stream is a difficult task for even the least complex gases. For gaseous mixtures of some complexity, such as air, very little reliable instrumentation is available.

One of the least complicated of species-concentration measurements is the determination of the degree of dissociation of a pure diatomic gas, or in a mixture of a diatomic gas with a noble gas [154]. For this measurement, some of the techniques which have had some success have involved interferometry, schlieren, selective absorption of radiation and similar spectroscopic techniques, mass spectrometry, electron spin resonance, and differential catalytic probes. Also, if the gas is in local thermodynamic equilibrium and temperature and electron density are known, each of the species concentrations may be determined from thermodynamic calculations. A measurement of absolute line intensities will also yield state populations if the transition probabilities are known. In general, however such considerations are not applicable to arc plasma work.

In applying interferometry to the measurement of dissociation in a diatomic gas, the specific refractivity of the atomic and molecular species and the total density of the gas must be known. Then by measuring the total refractive index of the gas by interferometry, the degree of dissociation of the gas may be determined [160]. Schlieren techniques have been applied but these yield only semi-quantitative results. For the use of techniques involving the selective absorption of radiation, the absorption coefficient of the different species must be determined, either through measurement or by calculation. Although time-of-flight mass-spectrometer measurements have shown promise, they have not been widely used. The electron-spin-resonance technique likewise has found little use because of the difficulty in its application.

Several investigators have attempted to apply differential catalytic heat-transfer probes to the determination of local atomic concentrations in high-speed or equilibrium streams of dissociated gases [161, 162]. These probes are essentially calorimeters with surfaces of known catalytic properties. By noting the difference in heat transfer to surfaces of different degrees of reactivity, a measure of the dissociation can be obtained. These probes also appear to yield only semiquantitative results.

The measurement of ionization and electron density has been accomplished with spectroscopy, Langmuir probes, and microwave interaction with the plasma gas, all of which were discussed earlier. Other techniques which have been proposed include electrical conductivity and induction effects, and optical and infrared interferometry.

Spectroscopic techniques are the most widely accepted methods for determining ionization and electron density in a flowing plasma. If the temperature of the gas has been measured by one of the spectroscopic techniques discussed earlier, then it is possible to calculate the electron density directly from equilibrium thermodynamic considerations, provided composition of the gas and all pertinent thermodynamic data are available [154]. However, the effective lowering of the ionization potential of the gas as a function of electron and ion density, and the effects of Coulomb correlation on the gaseous enthalpy, must be considered. Since the electron density determines almost completely the Stark width of a line, measured line profiles can be used to calculate electron densities, even though temperatures may not be known [163]. The electron density can also be determined from an absolute measurement of the emission due to the electron continuum in a spectral region or the measurement of the absolute intensity of an ionic line. In the latter case the transition probability must be known.

Conductivity meters which measure the electrical conductivity of the plasma have been used to determine electron density. One such instrument which measures the product of DC conductivity and gas velocity uses a sensing coil to determine the currents which are induced in the gas by primary exciting coils [164]. The DC conductivity in the plasma may be related to the electron density and the collision frequency. If the collision frequency and the velocity of the plasma are known, the electron density may be immediately derived from the meter reading. In another conductivity meter, operating on a slightly different principle, the resistive loading of a sensitive RF oscillator detector immersed in the plasma is used as a measure of the conductivity of the medium [165].

The phase change experienced by an electromagnetic wave passing through a plasma can be measured by an interferometer and used to determine the electron concentration in the gas. Such interferometric techniques have been used in the microwave region as well as in the far-infrared and optical regions [154]. Generally the microwave measurements are limited in size and electron densities which can be determined. Infrared techniques are capable of density determinations higher than those with microwaves but lower than can be determined in the optical regime.

7.8. Catalytic Effects

The condition of the high-temperature air as it comes in contact with a solid object has a pronounced effect on the rate of energy or heat transfer to that solid object. This is true with vehicles in actual flight and also with models in laboratory hyperthermal test facilities.

As an object passes through the earth's atmosphere at high speed, it experiences what is generally referred to as aerodynamic heating. This aerodynamic heating is not entirely a result of air friction, as is often implied, since the energy does not actually result from a shearing action in the air. As a high-velocity vehicle passes through the air a detached normal shock wave is set up in front of stagnation regions, such as the nose. As this shock wave passes through the ambient air, the properties of the air undergo a drastic and rapid change.

Since the air is compressed as the shock wave passes, the pressure, temperature and density increase. The normal component of the velocity of the air relative to the vehicle decreases from supersonic to subsonic. Although shock waves have finite thickness, usually comparable to the mean free path of the fluid molecules, they are generally considered as planes or very thin regions of discontinuity for engineering purposes. The fluid properties are treated as being discontinuous through these regions. An indication of the magnitude of the changes in fluid properties across a normal shock is evident from the data of Table 14. The data

Table 14. *Properties Across a Normal Shock Wave*
(Ref. 166)

Mach Number		Pressure	Density	Temperature
M_1	M_2	P_2/P_1	ϱ_2/ϱ_1	T_2/T_1
2	0.58	4.5	2.67	1.69
4	0.43	18.5	4.57	4.05
6	0.40	41.8	5.27	7.94
8	0.39	74.5	5.57	13.39
10	0.38	116.5	5.71	20.39

Subscripts 1 and 2 indicate properties preceding and following the normal shock, respectively.

are for a perfect gas having a 1.4 ratio of specific heats. As is evident from this, aerodynamic heating is not necessarily a result of friction, but rather is a case of convective heating from a high-pressure, high-temperature gas to the surface of a solid object.

When these large quantities of energy are imparted into a gas such as air, the amplitude of the vibrational motion of the atoms within the molecule becomes excessive. As this vibrational amplitude reaches a certain level the cohesive molecular bond is broken and the molecule is separated into individual atoms. This condition, characterized by the reactions:

$$N_2 + E_{D_N} \rightarrow N + N \tag{55}$$

and

$$O_2 + E_{D_o} \rightarrow O + O \tag{56}$$

is referred to as dissociation of the gas. The E_D quantities are the energy (in electron volts) required for dissociation of each major constituent, nitrogen and oxygen. If sufficient energy is then added to the dissociated atoms they will further become ionized. This is represented by:

$$N + E_{I_N} \rightarrow N^+ + e \tag{57}$$

and

$$O + E_{I_o} \rightarrow O^+ + e \tag{58}$$

where E_I is the energy required to completely remove a single electron from orbit in the atom. Imparting additional energy results in more electrons being removed from the atom. The additional increments of energy which must be added for second, third, and higher electron removal becomes higher and higher since the bonds of these electrons are progressively stronger.

Although the molecules of diatomic gases such as nitrogen or oxygen usually undergo dissociation prior to ionization, it cannot be assumed that a given quantity of the gas will be completely dissociated before any ionization takes place.

The effects of dissociation can be seen in the temperature of the gas. The temperature behind a normal shock at Mach 14 in air at 300° K is calculated by basic thermodynamic relationships to be in excess of 11,000° K. However, a more realistic prediction [167] of the temperature behind this shock is on the order of 6500° K, provided the air is given sufficient time to undergo dissociation and come to equilibrium. The reversal of the dissociation process, called recombination, is accompanied by a release of energy and likewise requires a finite period of time.

It is obvious that since the heating of the vehicle is primarily due to convection from the hot gas, the heat-transfer rate would be lower from a gas which remains dissociated than from one which recombines. The nature of the surface to which the heat is being transferred has a very pronounced effect on this heating rate from a dissociated gas. Some materials will act as catalysts to the recombination reaction; that is, they will increase the rate at which the gas molecules recombine. Other materials are not catalytic to the reaction but permit it to proceed at its normal rate. In the latter case, the flow over such a non-catalytic surface is essentially frozen in nature. With a catalytic surface the heat-transfer considerations become much more complex and must be treated as a case of a reacting boundary layer.

There is additional heating from a dissociated gas to a catalytic surface (as compared to a non-catalyst) due to two major effects. First, the energy which is released by the recombination of atoms, which now occurs very near the surface, is immediately available to be transferred to the surface. Therefore the boundary layer temperature is actually higher in the case of the catalytic surface. In addition there is a diffusion of atoms toward the surface and of recombined molecules away from the surface. The increased diffusion of atoms and molecules across the boundary layer results in an additional rate of energy transfer through the boundary layer from the free stream to the surface. This increased diffusion and corresponding transfer of energy across the boundary layer is the primary reason for greater heat transfer to a catalytic surface. With the non-catalytic surface the dissociated gas impinges on the surface and recombination is much slower. Therefore, the diffusion of atoms and molecules across the boundary layer is much less vigorous and the heat-transfer rate to the non-catalytic surface is lower. This is the essence of the catalytic effect of materials.

It is important to remember:

(1) That the heat flux through a boundary layer is strongly dependent on the diffusion of atoms and molecules within the boundary layer.
(2) That this diffusion is controlled by the rate at which the dissociated gas can undergo recombination, and this rate of recombination is increased by the presence of a catalytic surface.

Since the heat transfer through a boundary layer is strongly dependent on the recombination rate, it is obvious that for gases which recombine at a high rate the effect of a catalytic surface would be minimal. However, the recombination rates of both oxygen and nitrogen are slow and the catalytic effect can therefore be quite pronounced in air. The recombination rate of oxygen is considerably faster than that of nitrogen and has the strongest effect in heat transfer during flights through air.

The catalytic effect on the heat-transfer rate can be very significant. It has been shown [168] that by treating the non-catalytic wall as a frozen boundary layer and the catalytic wall as a simple first order reaction $(O+O \rightarrow O_2+E_D)$ within the boundary layer, a reduction of 50% in heat-transfer rate can be realized by using a non-catalytic surface.

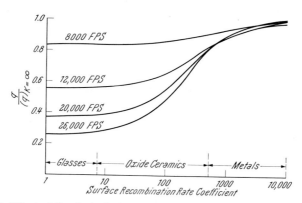

Fig. 73. Effect of Catalytic Coefficient of Materials on Heat Transfer Rate

Fig. 73 [169] shows the effect of surface recombination rate upon the stagnation-point heat transfer from dissociated air to an object with a spherical nose radius of 50 centimeters. The ordinate of this curve is an expression for the ratio of non-catalytic to catalytic heat transfer. These data were calculated for various velocities at conditions of level flight at 250,000 feet altitude and an assumed constant wall temperature of 700° K. It was assumed that the gas at the outer edge of the boundary layer behind the shock wave was in thermodynamic and dissociative equilibrium. The parameter of the abscissa represents a measure of catalytic reaction rate. On the bottom of this curve, superimposed on their approximate range of catalytic reaction rates, are three general classes of materials. The metals generally appear to be very good catalysts and the glasses very poor, with the oxide ceramics between these two extremes.

It is very important to remember that these calculations were made assuming level flight and assuming that the air behind the normal shock was completely (or nearly) dissociated. If an object enters the atmosphere at a relatively sharp angle, the density in the boundary layer increases very rapidly and there is insufficient time for the dissociation reaction to proceed to completion. This would mean that there would be only a small degree of dissociation of the air and the effect of the catalytic nature of the surface would be very slight. On the other hand, for objects entering the atmosphere at slight angles (i.e., gliding reentry from great altitudes and hypersonic velocities) the degree of dissociation would be quite high and the effect of a catalytic surface would be significant.

In the proper flight regimes, the non-catalytic properties of a material can be used to advantage in the design of a heat protection system. Thin oxide-ceramic coatings on the nose portions of lifting reentry vehicles have resulted in a decrease in substrate temperatures which were considerably greater than had been predicted from purely insulative effects. The same was true for a mixed-oxide ceramic coating applied to the frontal sections of a supersonic rocket sled. These applications have been discussed more fully in Chapter 4.

The gas which emerges from a plasma generator is likewise dissociated and ionized. The gas which passes through the arc column and forms the core of the plasma is in a highly excited state. In a jet which is stabilized on nitrogen this means that there are a considerable number of nitrogen molecules which are ionized. The oxygen which is added downstream to simulate air is also dissociated and ionized as it mixes with the very hot nitrogen. If the oxygen is injected through the arc itself, as it is in some units, an even greater degree of ionization will result.

This condition can be detrimental when using the plasma for flame spraying. Heat transfer from the highly ionized gas is lower to oxide ceramic particles than to metallic powders. This, coupled with the generally high melting points, specific heats, and low thermal conductivities of ceramics, can make spraying of these materials difficult. When spraying mixtures of metals and ceramics, the problem of different rates of heat transfer to the different powders can be particularly troublesome. On the other hand, when conducting materials evaluations the conditions of dissociation and ionization of the stream are considered as advantageous. It enhances the simulation of reentry conditions. A material's performance will be affected by its catalytic nature in both the laboratory evaluation and actual flight. A catalytic material will be subjected to a higher heat flux in each case than a non-catalytic material under identical conditions.

The rate of recombination on the surface also has an effect on diagnostic instrumentation which is used in an arc-plasma-generated stream. The catalytic effect can be especielly strong with calorimeters. The majority of calorimeter sensing surfaces are metallic, usually copper, and as such are strong catalysts to recombination. If the surfaces of these probes are permitted to become contaminated or if an oxide layer forms, their catalytic nature can change appreciably.

It must also be remembered that the material which is being evaluated may be a non-catalyst and will therefore not be subjected to the same conditions as was the calorimeter. The magnitude of this effect, of course, depends entirely on the degree of dissociation and ionization in the particular plasma effluent being used. For example, consider a non-catalytic material which is to be evaluated in two arc plasma facilities which produce vastly differing degrees of dissociation and ionization. If the same calorimeter (with a catalytic sensing surface) were used to establish the test conditions, the non-catalytic test sample would actually be subjected to more severe heating from the plasma stream which has the lower degree of dissociation and ionization. Although a quantitative measure of these catalytic effects is extremely difficult to obtain, the effects are nevertheless real and should not be neglected when analyzing a material's performance.

7.9. Material Response

The performance of a material is generally evaluated by its behavior during exposure and the effect produced by the exposure. Qualitative observations and quantitative measurements made before, during, and after a model has been exposed are helpful in determining its performance. The parameters of interest and the techniques used to obtain the desired data vary with the type of material being evaluated and the test being performed. However, there are some parameters, such as temperature response, which are generally of interest regardless of the test or material type.

The methods for monitoring material temperatures can be divided into two types, those incorporating direct-contact sensors and those in which the transducer makes no physical contact with the specimen. Internal or back-surface temperatures are usually sufficiently low to permit use of the direct attachment of thermocouples, while exposed-surface temperatures require the use of non-contact radiation techniques either because of the high temperature involved or the dynamic physical conditions (such as ablation) which exist during test.

Thermocouples are widely used to determine internal temperature measurements on in-flight vehicles as well as on specimens or models in laboratory test facilities. Because of the importance of these measurements to the analysis of a material's performance and the ultimate design of operational components, it is essential that care be exercised to minimize the sources of errors in obtaining these measurements. In addition to the normal precautions to be exercised when using thermocouples (such as welding junctions, maintaining cleanliness of the junction areas and lead wire, proper selection of thermocouple type for the temperature range to be measured and proper use of instrumentation for reading the emf output) care should also be taken to minimize the errors which occur from several additional sources [170]. These include thermocouple-location accuracy, electrical shorting through electrically conductive products which may be formed during exposure, and, most importantly, thermal disturbances which result from the very presence of the sensor.

Accurate knowledge of the precise location of the thermocouple junction is essential, to assure reproducibility of data from specimen to specimen and for accurate use of the data for design purposes. This is obviously more critical in laboratory-size models because a relatively small error in location can represent a significant percentage of the overall dimensions. Whenever possible, the exact location of each hot junction should be verified by radiographic techniques prior to exposure. X-rays taken in two planes can be used but care should be taken to correct for parallax.

The carbonaceous char layer which is formed by the decomposition of most organic materials during ablation is highly conductive. Shorting of thermocouples by this char can result in temperature errors up to 200° F [170]. Care should therefore be taken to protect the lead wires from such electrical shorting by using ceramic coatings or protective tubing whenever possible.

It would be ideal to measure the internal temperature, with no foreign substance present to create a disturbance in the thermal conditions and flow of heat within the body. Since this is not physically possible, precautions should be taken to minimize the disturbance introduced by the presence of the sensing element. The bead at the hot junction should be as small as possible, from one-and-one-half to two times the wire diameter. This is necessary because large beads result in erroneously low temperature measurements because of the heat-sink effect of the bead. The error is most critical in dynamic conditions and with materials whose thermal conductivity differs significantly from that of the thermocouple.

There is another effect, which is especially prevalant when the conductivity of the sample differs appreciably from that of the thermocouple, which can easily result in errors of several hundred degrees [171]. This error results from heat being conducted away from the vicinity of the junction along the thermocouple lead wires at a higher rate than the rate at which it is transferred to the site. When this occurs, the temperature in the vicinity of the hot junction remains considerably lower than it would be if the thermocouple were not present. This occurs even during conditions of steady-state operation. This source of error can be minimized by installing the sensor so that the lead wires emanating from the bead extend for some distance in an isothermal plane which includes the hot junction. It is recommended that these lead wires extend a minimum of 25 wire diameters from the bead [170].

When the temperature of the body is too high or where the physical conditions preclude the use of thermocouples, it is usually necessary to resort to remote optical means to determine temperatures. All bodies, by virtue of their temperature, emit radiant energy from their surfaces at the expense of the heat energy which they contain. The rate at which this radiant energy is emitted is dependent upon the absolute temperature of the body and upon its emittance.

There are a number of pyrometers which can be used to measure and record the radiation or brightness temperature of a body. Most of these pyrometers are based on the Stefan-Boltzmann law which states that the total radiation per unit area and unit time from a blackbody (emittance equal to unity) is proportional to the fourth power of its absolute temperature. That is

$$\Phi = \sigma T^4 \tag{59}$$

where Φ is the radiant energy flux, σ is the Stefan-Boltzmann constant, and T is the absolute temperature of the radiating body. If the conditions are non-blackbody, this relationship is changed by introducing an emissivity term, ε, on the right hand side of Equation 59. The radiant flux from a non-blackbody is therefore proportional to σ, T^4, and ε.

The instruments which are used to determine temperatures optically can be classified roughly into two categories; chromatic total-radiation sensors and intensity meters [172]. The chromatic instrument make use of one or more colored filters which permit narrow bands of wave lengths of light from the luminous hot body to pass to the eye or to an inanimate sensor. Usually the energy falling on the sensor is related to a reference standard to obtain an apparent temperature. Typical monochromatic optical pyrometers such as the Leeds and Northrup or Pyro instruments are relatively inexpensive but are manually operated. Dichromatic pyrometers such as the Shawmeter or Coloratio operate and record automatically but are more expensive. Polychromatic instruments having three or more filters have been developed but are not widely used.

Total-radiation pyrometers measure the apparent temperature of a body by focusing the energy which is radiated from the hot area onto an absorbing area or sensor within the instrument. The sensing area is a single thermocouple, a thermopile, a bolometer, or similar heat-sensitive cell. The output of this sensing cell can be read directly from a meter or recorded. With these units it is necessary to have a window of fused silica, fluorite, glass, etc., to protect the sensing element. Such windows absorb various wave lengths of radiated energy and thus introduce absorption errors. The instruments can therefore no longer be truly called total-radiation pyrometers. Typical total-radiation pyrometers include Ray-O-Tube and Land Eye.

Intensity meters simply measure the intensity of filtered segments of radiation, using an intensity-sensitive sensor, and relate its energy output to the temperature of the radiant source. There is a variety of these instruments available which operate in different ranges of wavelengths, including the infrared. The infrared units offer the advantage of accurate determinations at low temperatures (room temperature or below).

If these pyrometers were to be used to determine only the temperatures of blackbodies, which are ideal emitters, or even gray bodies, which have an emissivity which is nearly constant for all wavelengths and temperatures, there would be no problem. Unfortunately all real bodies are something other than blackbodies and very few materials behave like gray bodies. Most materials emit radiant energy non-uniformily at various wavelengths. Furthermore, when dealing with the surfaces of materials undergoing test in an arc plasma facility the material is even less ideal. Surfaces are certainly somewhat less than optically flat, the finish is generally less than mirror polished and the surface is not virgin, but contains some film, oxide, or other substance. The body is usually not in a static condition but supports some process of chemical reaction, attrition, or degradation. In other words it is in a dynamic or changing state. Therefore the term emissivity, which by definition is a function of an ideal body, no longer applies. Instead the term emittance is used to describe this less-than-ideal condition. Emittance is the ratio of the energy reradiated by a non-ideal body to the incident energy from a radiant source. Emissivity is the lowest limiting value of emittance. Emittance is not a constant characteristic of the material. It refers only to the body or material surface at the time of evaluation and under specific environmental conditions. It is essentially an instantaneous value, not necessarily reproducible. Therefore it should not be cited as a handbook value, but must be qualified by the test and sample conditions.

Any temperature measurement of a hot body made by optical means is only an indicated or apparent temperature which must be corrected for the emittance of the body before a true temperature may be obtained. If the true temperature of a non-ideal body needs to be known, then the emittance factor must be applied to correct the apparent temperatures as determined by the temperature-sensing instrument.

Emissivity values as they are reported in textbooks, handbooks, and most of the technical literature are actually invalid for use in converting apparent pyrometer temperatures to true temperatures. Uncertainty exists as to whether the data were reported as emissivities or emittances, since in the great majority of instances no information is given about the condition of the specimens or the nature of the test procedures. If the data reported are emissivity values then they are not usable directly since they reflected an ideal emitter, which is not the condition of a sample under test conditions. For that matter even if they are emittances they cannot be used, because they are almost certainly not representative of the experiment at hand. The problem, of course, is to accurately determine the emittance of a body while it is undergoing some other type of thermal evaluation, such as a test in an arc plasma facility.

References

1. HELLUND, E. J.: The Plasma State. New York: Reinhold 1961.
2. GIANNINI, G. M.: The Plasma Jet. Sci. Amer., **107** (2) 80—88 (1957).
3. DENNIS, P. R., C. R. SMITH, D. W. GATES, and J. B. BOND: Plasma Jet Technology. NASA SP-5033, National Aeronautics and Space Administration, 1965.
4. THORPE, M. L.: The Plasma Jet and Its Uses. Res./Develop., **11** (1) 5 (1960).
5. SMITH, D. M.: Plasma Spraying of Refractory Materials. Gen. Motors Eng. J., 2nd quarter, 1963.
6. SMITH, H. E.: Development and Investigation of an Arc-Plasma Material Evaluation Facility. Wright-Patterson A.F.B., ASD-TDR-62, U.S. Air Force, 1962.
7. PASK, J. A.: Ceramic Processing, NAS, No. 1576. National Academy of Science, 1968.
8. BROWN, S. D., et al.: Critical Evaluation of Ceramic Processing at Subconventional Temperatures. Wright-Patterson A.F.B., AFML-TR-67-194, U.S. Air Force 1967.
9. INGHAM, H. S., and A. P. SHEPARD: Flame Spray Handbook, **I**, Metallizing Handbook — Wire Process, **II**, Powder Process, **III**, Plasma Flame Process, Westbury, Long Island: Metallizing Engineering Co., Inc., 1965.
10. BROPHY, J. H., H. W. HAYDEN, K. G. KREIDER, and J. WULFF: Activated Sintering of Pressed Tungsten Powders and Plasma Jet Sprayed Tungsten Deposits. Bureau of Naval Weapons Contract NOas 61-0326 O C, U.S. Navy, 1961.
11. ANONYMOUS: Shop Notes for Material Spray Application with Plasmadyne S Series Hand Spray Guns. Santa Ana, Calif., Plasmadyne Corp. (No Date).
12. LEEDS, D. H.: Interface Bonding Studies and a New Plasma-Arc Spray Gun, Accomplishment and Plans. Summary of the Fifth Refractory Composites Working Group Meeting, HJELM, L. N., Wright-Patterson A.F.B., ASD-RDR-63-96, U.S. Air Force, 1961.
13. JOHNSON, R. H., and W. H. WHEILDON: Faster Plasma Coating. Mater. Design Eng., **56** (6) 16—17 (1962).
14. WHEILDON, W. M.: The Role of Flame-Sprayed Ceramic Coatings as Materials for Space Technology and the Systems for Applications. Presented at the American Institute of Chemical Engineering Meeting, March, 1963.
15. LEVINSTEIN, M. A., A. EISENLOHR, and B. E. KRAMER: Properties of Plasma Sprayed Materials. Welding Res. Supplement, **1**, 8 S—13 S (1961).
16. LEVY, M.: Trip Report to the III International Metallization Conference. Madrid, Spain. Watertown Arsenal Laboratories, U.S. Army (1962).
17. MASH, D. R., N. E. WARE, and D. L. WALKER: Process Variables in Plasma-Jet Spraying. J. Metals, **13**, (7) 473—478 (1961).
18. GRISAFFE, S. J., and W. A. SPITZIG: Preliminary Investigation of Particle-Substrate Bonding of Plasma Sprayed Materials. NASA TN-D-1705, National Aeronautics and Space Administration, 1963.
19. LEVINSTEIN, M. A.: Plasma Spraying-State of the Art. AF 33 (616) 6376 Task 7381 U.S. Air Force, 1961.
20. MARYNOWSKI, C. W., F. A. HALDEN, and E. P. FARLEY: Variables in Plasma Spraying. Electrochem. Tech., **3** (3—4) 105—115 (1965).

21. ENGELKE, J. L.: Heat Transfer to Particles in the Plasma Flame. Presented at the American Institute of Chemical Engineers' meeting Los Angeles, California, January, 1962.

22. HECHT, N. L.: Plasma Sprayed Ceramic-Metal Composite Coatings. Presented at the 68th Annual Meeting of the American Ceramic Society, May, 1966.

23. WURST, J. C., J. A. CHERRY, D. A. GERDEMAN, and N. L. HECHT: The Evaluation of Materials Systems for High Temperature Aerospace Applications. Wright-Patterson A.F.B., AFML-TR-65-339 Part I, U.S. Air Force, 1966.

24. SMITH, G. D.: Heat Transfer Characteristics of Materials in Plasma Flames. Presented at the Fourth Refractory Composites Working Group Meeting, November, 1960, preprint.

25. PLUNKETT, J. D.: NASA Contribution to the Technology of Inorganic Coatings. NASA SP-5014 Technology Survey, National Aeronautics and Space Administration, 1964.

26. MOORE, D. G., et al.: Studies of the Particle-Impact Process for Applying Ceramic and Cement Coating. ARL-59, AD 266381, U.S. Air Force, 1961.

27. THOMPSON, V. S., and O. J. WHITTEMORE: Structural Changes on Reheating Plasma-Sprayed Alumina. Amer. Ceram. Soc. Bull., **47** (7) 261—289 (1968).

28. BLITON, J. L., and H. L. RECHTER: The Design of a Flame-Spray Coating. Solid Bodies Society of Aerospace Materials and Process Engineers, North Hollywood, Calif., Western Periodicals Company, 1961.

29. MATTING, M. A., and H. D. STEFFENS: Haftung und Schichtaufbau beim Lichtbogen und Flammspritzen. Metal, **17** (6) 583—593, (9) 905—922 and (12) 1212—1227 (1963).

30. BLITON, J. L., and H. L. RECHTER: Determination of Physical Properties of Flame-Sprayed Ceramic Coatings. Amer. Ceram. Soc. Bull., **40** (11) 683—688 (1961).

31. HUFFADINE, J. B., and A. G. THOMAS: Flame Spraying as a Method of Fabricating Dense Bodies of Alumina. Powder Met., **7** (14) 290—299 (1964).

32. LEEDS, D. H.: Materials and Structure Program. Aerospace Corp., Report No. TDR-930, DCAS TDR-62-47, U.S. Air Force, 1962.

33. STRAUSS, I.: Final Report on Development of Composite Rocket Nozzles. Special Projects Office TR 61-109. 8, NOw 61-0479-C, U.S. Navy (1961).

34. LASZLO, T. S.: Mechanical Adherence of Flame-Sprayed Coatings. Amer. Ceram. Soc. Bull., **40** (12) 751—755 (1961).

35. GRISAFFE, S. J.: Analysis of Shear Bond Strength of Plasma Sprayed Alumina Coatings on Stainless Steel. NASA-TN-D-3113, National Aeronautics and Space Administration, 1965.

36. NEWCOMER, R.: Precoats for Adhesion of Sprayed Aluminium Coatings. McDonnell Aircraft Corp. Report A 241—AF 33 (657)-11215, U.S. Air Force, 1963.

37. BLITON, J. L.: Plasma-Sprayed Oxide and Vapor-Deposited Nitride Coatings on Tungsten as a Means of Achieving Oxidation Protection. Summary of the Seventh Refractory Composites Working Group Meeting, I, HJELM, L. N., D. R., JAMES, Wright-Patterson A.F.B., RTD-TDR 63-4131, U.S. Air Force, 203—208, 1963.

38. SPITZIG, W. A., and S. F. GRISAFFE: Metallurgical Bonding of Plasma-Sprayed Tungsten and a Stainless Steel Substrate. NASA TN-D-2461, National Aeronautics and Space Administration, 1964.

39. LONGO, F. N.: Metallurgy of Flame Sprayed Nickel Aluminide Coatings. Welding Res. Supplement, **2**, 66 S—69 S (1966).

40. ALSOP, R. T., et al.: The Adhesion of Sprayed Molybdenum. Metallurgia, **63** (4) 125—131 (1961).

41. HOWINK, R., and G. SALOMON: Adhesion and Adhesives, **I**, Amsterdam: Elsevier Publishers, 1965.

42. Spitzig, W. A., and S. J. Grisaffe: Metallurgical Bonding of Plasma-Sprayed Tungsten on a Hot Molybdenum Substrate. NASA TN-D-2510, National Aeronautics and Space Administration, 1964.

43. Meyer, H., and A. Dietzel: Das Flammspritzen von keramischen Überzügen. Ber. Deut. Keram. Ges. **37.**, (4) 136—141 (1960).

44. Bliton, J. L., and R. Havell: Physical Properties of Flame-Sprayed Ceramic Coatings Part II BaTiO$_3$. Amer. Ceram. Soc. Bull., **41** (11) 762—767 (1962).

45. Roller, D.: Summary of the Second High Temperature Inorganic Refractory Coating Working Group Meeting. Wright-Patterson A.F.B., WADC TR-59-415-915, U.S. Air Force, 1959.

46. Bliton, J. L., Y. Harada, and Richter: Flame Sprayed Zirconia Films for Fuel Cell Components. Amer. Ceram. Soc. Bull., **42** (1) 6—9 (1963).

47. Brophy, J. H., H. R. Herdeklang, K. G. Kreider, and J. Wulff: Activated Sintering of Pressed Tungsten Powders and Plasma Jet Sprayed Tungsten Deposits. Bureau of Naval Weapons, Contract NOas 59-6264c, U.S. Navy, 1960.

48. Landingham, R. L., et al.: Plasma-Jet Coatings of Tungsten on Steel. AD 625800, University of Arizona, 1962.

49. Eisenlohr, A.: Arc Plasma Sprayed Tungsten as an Engineering Materials. Summary of the Seventh Refractory Composites Working Group Meeting **III**, Hjelm, L. N., D. R. James. Wright-Patterson A.F.B., RTD-TDR 63-4131, U.S. Air Force, 1963.

50. Halkias, J. E., H. R. Thornton, and J. E. Burroughs: Plasma Arc Deposition for Aerospace Applications. Plating, **52** (1), 44—54 (1965).

51. Wheildon, W. M.: Flame Sprayed Ceramic Coating Survey and State of the Art. Presented at the 69th Annual Meeting of the American Society for Testing Materials, 1966.

52. Ingham, H. S., and A. P. Shepard: Evaluation Methods and Equipment for Flame-Sprayed Coatings. Research Laboratory Report #106, Westbury, Long Island: Metco, Inc., 1963.

53. Hjelm, L. N.: Minutes of Flame-Spray Groups. Internal Report ASTM C-22, 1966.

54. Polena, M.: Flame Sprayed Ceramic Coating Techniques. Thiokol Reaction Motors Division Report RMD-9368 F, Denville N. J., 1964.

55. Huminik, J.: High-Temperature Inorganic Coatings. New York: Reinhold, 1963.

56. Pirogov, Y. S., R. M. Brown, and A. L. Friedberg: Electrical Properties of Al$_2$O$_3$-Nickel Metal Multilayer Flame-Sprayed Coatings. Amer. Ceram. Soc. Bull., **45** (12) 1071—1074 (1966).

57. Galli, J. R., G. I. Wheeler, G. H. Clampitt, D. E. German, and R. B. Johnson: Development and Evaluation of Rocket Blast and Rain Erosion Resistant Composite Coatings Produced by Flame Spraying Techniques. Wright-Patterson A.F.B., WADC TR 58-493, U.S. Air Force, 1959.

58. Liebert, C. H.: Spectral Emittance of Aluminium Oxide and Zinc Oxide on Opaque Substrates. NASA TN D-3115, National Aeronautics and Space Administration, 1965.

59. Meyer, D. H.: Über das Flammspritzen von Aluminiumoxyd. Werkstoffe und Korrosion, **11** (10) 601—616 (1960).

60. Grenis, A. F., and A. P. Levitt: Infrared Radiation of Solid Refractory Materials. Amer. Ceram. Soc. Bull., **44** (11) 901—906 (1965).

61. Hayes, R. J., and W. H. Atkinson: Thermal Emittance of Materials for Spacecraft Radiator Coatings. Amer. Ceram. Soc. Bull., **43** (9) 616—621 (1964).

62. Unger, R.: Fourth Quarterly Progress Report. U.S. Air Force Contract AF 33 (616) 7323, Plasmakote Corp., 1961.

63. Howe, J. E., and W. C. Riley: Ceramics for Advanced Technologies. Chapter 7, New York: John Wiley and Son, 1965.

64. McCullum, D. E., and N. L. Hecht: Plasma Sprayed Coatings for Thermal Protection of Rocket Sled Components. Engineering Test Memorandum #12, University of Dayton, U.S. Air Force Contract AF 33615-1312, 1966.

65. Firth, K. E., and W. E. Lawrie: Ultrasonic Methods for Nondestructive Evaluation of Ceramic Coating. Wright-Patterson A.F.B., WADD TR-61-91 Part II, U.S. Air Force, 1962.

66. Baumanis, A. M., and W. E. Lawrie: Ultrasonic Methods for Nondestructive Evaluation of Ceramic Coating. Wright-Patterson A.F.B., WADD TR-61-91 Part III, U.S. Air Force, 1963.

67. Lawrie, W. E., and M. D. Oestreich: Nondestructive Methods for the Evaluation of Ceramic Coatings. Wright-Patterson A.F.B., WADC TR-61-91, Part IV, U.S. Air Force, 1964.

68. Haines, K. A., and B. P. Hildebrand: Interferometric Measurements on Diffuse Surfaces by Holographic Techniques. Trans. on Instr. Meas., 15, (4) 149—161 (1966).

69. Hildebrand, B. P., and K. A. Haines: Interferometric Measurements Using the Wavefront Reconstruction Technique. Appl. Opt. 5., (1) 172—173 (1966).

70. Sollid, J. E.: Holographic Interferometry Applied to Measurements of Small Static Displacements of Diffusely Reflecting Surfaces. Appl. Opt., 8, (8) 1587—1595 (1966).

71. Hecht, N. L.: Refractory Coatings Development at the University of Dayton Research Institute. Presented at the 17th Refractory Composite Working Group Meeting, June, 1970.

72. Hauser, H. H. (ed.): Modern Materials, 2, p. 63, New York: Academic Press, 1960.

73. Moore, D. G., W. D. Hayes, and A. W. Crigler: Velocity Measurements of Flame-Sprayed Aluminium Oxide Particles. Phase I # 33 (616) 58—19 Project # 88-7022, U.S. Air Force, 1959.

74. Mock, J. A.: Flame Sprayed Coatings. Mater. Design Eng., 63 (2) 89—104 (1966).

75. Fabian, R. J.: What's New in Coatings and Finishes. Mater. Design Eng., 57 (4) 109 to 116 (1963).

76. Palermo, J. R., and C. C. Polter: Plasma Spraying Present and Future. Lebanon, N. H., Thermal Dynamics Corp. (No Date).

77. Fairlie, J.: Plasma for Cutting, Welding, Coating. Welding Eng., 47 (11) 41—44 (1962).

78. Moss, A. R., and W. J. Young: The Role of Arc-Plasma in Metallurgy. Powder Met., 7 (14) 261—289 (1964).

79. Ulrick, D. T., and E. J. Smoke: Devitrified Barium Titanate Dielectrics. J. Amer. Ceram. Soc., 49 (4) 210—215 (1966).

80. Wheildon, W. M.: Properties of Thermal Sprayed Zirconate Coatings. Presented at the 13th Annual Meeting of the Refractory Composites Working Group, July, 1967.

81. McGeary, T. C., and J. M. Koffskey: Engineering Applications for Flame Plating. Metal Prog., 87 (1) 80—86 (1965).

82. Sternkopf, J., and G. Ludke: Über das Flammspritzen von Keramikschicht. Die Technik, 19 (6) 398—402 (1964).

83. Bortz, S. A.: Four Short Review Sections From Two Current Major Programs at IITRI, Summary of the Ninth Refractory Composites Working Group Meeting, L. N. Hjelm, D. James, and E. Beardsley: Wright-Patterson A.F.B., AFML-TR-64-398, U.S. Air Force, 825—861, 1965.

84. Merry, J. D., and C. H. Vondrocek: Three Uses of Flame-Sprayed Al_2O_3. Mater. Design Eng., 57 (2) 72 (1963).

85. Shepard, A. P.: Sealer Extends Rust-Free Life for Metallized Structures. Iron Age, 191 (15) 69—71 (1963).

86. Anonymous: Mounting Sensing Elements with Rokide Ceramic Coatings. Worcester, Mass.: Norton Co., 1963.

87. AULT, N. N., and L. H. MILLIGAN: Alumina Radomes by Flame-Spray Process. Amer. Ceram. Soc. Bull., **38** (11) 661—664 (1959).

88. ALLEN, A. C.: Plasma-Jet A New Tool for Ceramics. Ceram. Ind., **82** (4) 112—114 (1964).

89. ANONYMOUS: Spray Coatings. Welding Eng., **50** (7) 37—46 (1965).

90. OTTE, H. M., and S. R. LOOKE: Material Science Research **2**, New York: Plenum Press, 1965.

91. ANONYMOUS: Protective Coatings for Refractory Metals in Rocket Engines. Contract No. NAS 7—113, Summary Report, National Aeronautics and Space Administration, 1963.

92. LEVY, M.: Refractory Coating Research and Development, Watertown Arsenal Laboratories. Summary of the Fifth Refractory Composites Working Group Meeting, HJELM, L. N. Wright-Patterson A.F.B., ASD-TDR-63-96, U.S. Air Force, 1961.

93. BLITON, J. L., and S. W. BRADSTREET: Pertinent Activities in Refractory Composites. Ibid.

94. VASILOV, T., and G. HARRIS: Impervious Flame-Sprayed Ceramic Coatings. Amer. Ceram. Soc. Bull., **41** (1) 14—17 (1962).

95. BLITON, J. L., and S. A. BORTZ: Ceramic Coatings for Cementitious and Metallic Surfaces. IITRI-B8009-7, Illinois Institute of Technology (1964).

96. PASTERICK, N. R., and G. W. FISHER: Plasma Metallizing a Compressor Case. Amer. Machinist/Metalwork Manufac., **167**, (22) 87 (1963).

97. ROLLER, D.: Summary of the First High Temperature Inorganic Refractory Coating Meeting. Wright-Patterson A.F.B., WCLT-TM-58-139, U.S. Air Force, 1958.

98. BORTZ, S. A.: A Review of Current Refractory Composite Research in Coatings Meeting. Ibid.

99. KNANISHU, J.: Galvanic Protection by Metal Spray Method. Report #63 1491 Rock Island Arsenal, U.S. Army, 1963.

100. ANONYMOUS: Flame Spraying Multiplies Life of Worn Parts for Brick Making Machine. Westbury, Long Island, Metallizing Engineering Co. Inc. (No Date).

101. LEVY, M.: Refractory Coating Research and Development at U.S. Materials Research Agency. Summary of the Seventh Refractory Composites Working Group Meeting, **II**, HJELM, L. N., and L. N. JAMES. Wright-Patterson A.F.B., RTD-TDR-63-4131, U.S. Air Force (450—467) (1963).

102. HERRON, R. H.: Research at Bendix Aviation Corp. Summary of Second High-Temperature Inorganic Refractory Coatings Working Group Meeting, ROLLER, D., WADC TR-59-415, U.S. Air Force, 41—43, 1959.

103. STETSON, A. R.: Tungsten Wire Reinforced Plasma-Arc Sprayed Tungsten. Summary of the Sixth Refractory Composites Working Group Meeting, ASD TDR-63-610 **I**, HJELM, L. N., U.S. Air Force, 187—197, 1962.

104. DORSEY, J. J.: Development and Evaluation Services on Ceramic Materials and Wall Composites for High-Temperature Radome Shapes. WADC TR-57-665, U.S. Air Force, 1958.

105. HAYES, H. A.: An Investigation of the Feasibility of Forming Alloy Coatings with a Plasma Jet. Naval Weapons Report 7617 NOTS TD 2616, U.S. Navy, 1961.

106. LEVY, M., and A. P. LEVITT: Application of Flame-Sprayed Coatings at Watertown Arsenal. Technical Report NOWAL TR 3711/1 Watertown Arsenal, U.S. Army, 1961.

107. LEVY, M., and A. P. LEVITT: Flame-Sprayed Metallic and Ceramic Coatings for Army Applications. AMRAMS 64-01, Army Materials Research Agency, U.S. Army, 1964.

108. KREMITH, R. D., *et al.*: Solid Lubricant Coatings Applied by the Plasma Spray Process. Amer. Ceram. Soc. Bull., **47**, (9) 813—818 (1968).

109. KOCK, W. E.: Lasers and Holography, Garden City, New York: Doubleday, 1969.

110. HOERCHER, H. E.: Summary of Plasma Facilities in the United States. Avco Corp., August, 1966.
111. SCHWARTZ, M. A.: Proposed Definition of Ablation. ASTM Committee E-21, Section 3 Letter Ballot, March, 1967.
112. KATSIKAS, C. J., et al.: Ablation Handbook Entry Materials Data and Design. AFML TR-66-262, U.S. Air Force, 1966.
113. WURST, J. C., and D. A. GERDEMAN: Arc Heater Screening of Ablative Plastics. AFML TR-65-110, U.S. Air Force, 1965.
114. SCHMIDT, D. L., and H. S. SCHWARTZ: Evaluation Methods for Ablative Plastics. Soc. Plastics Engr. Trans. 3., No. 4 (1963).
115. CAMPBELL, I. E.: High-Temperature Technology. New York: John Wiley and Sons, 1965.
116. KINGERY, W. D.: Introduction to Ceramics. New York: John Wiley and Sons, 1960.
117. BUESSEM, W. R.: Thermal Shock Testing. J. Amer. Ceram. Soc., 38, No. 1 (1965).
118. HAUSER, H. H.: Coatings of High Temperature Materials. New York: Plenum Press, 1966.
119. WURST, J. C., J. A. CHERRY, D. A. GERDEMAN, and N. L. HECHT: The High-Temperature Evaluation of Aerospace Materials. AFML-TR-66-308, U.S. Air Force, 1966.
120. WURST, J. C.: The Development of a Standardized Screening Test for High Temperature Materials. Summary of the Sixth Refractory Composites Working Group Meeting, ASD-TR-63-610, 2, U.S. Air Force, 1962.
121. KAMIN, J. I.: Aerodynamics Simulation by High Spin Rate in an Arc-Plasma Environment. Cincinnati Testing Laboratories, July, 1964.
122. ROBBINS, D. L.: Thermal Erosion of Ablative Materials. ASD-TR-61-307, U.S. Air Force, 1962.
123. WURST, J. C., and D. A. GERDEMAN: Screening of Candidate Rocket Nozzle Materials. Summary of the Tenth Refractory Composites Working Group Meeting, AFML-TR-65-207, U.S. Air Force, 1965.
124. HOERCHER, H. E.: Proposed Method for Measuring Plasma-Arc Gas Enthalpy by Energy Balance. ASTM Committee E-21, Section 3 Letter Ballot, February, 1967.
125. DAWES, L.: Electrical Engineering, I. (Chap. V) and II. (Chap. IV). New York: McGraw Hill Book Co., 1937.
126. Instr. Soc. Amer. Transducer Compendium, New York: Plenum Press, 1963.
127. Book of ASTM Standards: General Testing Methods; Statistical Methods; Appearance Tests; Temperature Measurement. ASTM Standards, Part 30 (1966).
128. WINOVICH, W.: On the Equilibrium Sonic-Flow Method for Evaluating Electric-Arc Air-Heater Performance. NASA TN D-2132, National Aeronautics and Space Administration, 1964.
129. LIEPMAM, H. W., and A. E. PUCKETT: Introduction to Aerodynamics of Compressible Fluid, New York: John Wiley and Sons, 1947.
130. HALLBACH, C. R.: Pneumatic Pyrometry. The Marquardt Corp., October, 1962.
131. GREY, J., P. F. JACOBS, and M. P. SHERMAN: Calorimetric Probe for the Measurement of Extremely High Temperature. Rev. Sci. Instr., 33, No. 7 (1962).
132. GREY, J.: Enthalpy Probes for Arc-Plasmas-Status Review. Presented at ASTM Committee E-21, Section 3 Meeting, April, 1966.
133. GREY, J.: Enthalpy Probes for Arc-Plasmas-Second Status Review. Presented at ASTM Committee E-21, Section 3 Meeting, May 3, 1967.
134. HASS, F. C.: An Evaporating Film Calorimetric Enthalpy Probe. ARL 63-47, U.S. Air Force, 1963.
135. VASSALLO, F. A.: Miniature Enthalpy Probes for High-Temperature Gas Streams. ARL 66-011, U.S. Air Force, 1966.

136. GREY, J.: Thermodynamic Methods of High-Temperature Measurements. Instr. Soc. Amer. Trans., **4**, No. 2, 1965.

137. LEES, L.: Laminar Heat Transfer Over Blunt-Nosed Bodies at Hypersonic Flight Speeds. Jet Propulsion, **26**, No. 14 (1956).

138. FAY, J. A., and F. R. RIDDELL: Theory of Stagnation Point Heat Transfer in Dissociated Air. J. Aeron. Sci., **25**, No. 2 (1958).

139. GOULARD, R. J.: On Catalytic Recombination Rates in Hypersonic Stagnation Heat Transfer. Jet Propulsion, **28**, No. 11 (1958).

140. SCALA, S. M.: Hypersonic Heat Transfer to Surfaces Having Finite Catalytic Efficiency. RM No. 4, Aerophysics Laboratory, July, 1957.

141. SIBULKIN, M.: Heat Transfer Near the Forward Stagnation Point of a Body of Revolution. J. Aeron. Sci., **19**, No. 8 (1952).

142. ROSNER, D. E.: Similitude Treatment of Hypersonic Stagnation Heat Transfer. Amer. Rocket Soc. J., **29**, No. 2 (1959).

143. FAY, J. A., and N. H. KEMP: Theory of Stagnation Point Heat Transfer in a Partially Ionized Diatomic Gas. Presented at IAS Annual Meeting, New York, January, 1963.

144. ROSNER, D. E.: Effects of Diffusion and Chemical Reaction in Convective Heat Transfer. Amer. Rocket Soc. J., **30**, No. 1 (1960).

145. TODD, J. P.: Proposed Method for Measuring Heat Flux Using a Thermal Capacitance (Slug) Calorimeter. ASTM Committee E-21, Section 3 Letter Ballot, March, 1970.

146. BIRD, R. B., W. E. STEWART, and E. N. LIGHTFOOT: Transport Phenomena. New York: John Wiley and Sons, 1960.

147. STEMPEL, F. C., and D. L. RALL: Applications and Advancements in the Field of Direct Heat Transfer Measurements. Presented at ISA Conference, September, 1963.

148. THOMPSON, R. E.: Proposed Method of Measuring Heat Flux Using a Multiple Wafer Calorimeter. ASTM Committee E-21, Section 3 Letter Ballot, September, 1968.

149. KENNEDY, W. S., C. A. POWERS, R. A. RINDAL, and K. A. GREEN: AFFDL 50 MW RENT Facility Calibration. First QPR, Aerotherm Corp., February, 1970.

150. BECK, J. V., and H. HURWICZ: Effect of Thermocouple Cavity on Heat Sink Temperature. J. Heat Transfer, February, 1960.

151. SHERMAN, F. S., and L. TALBOT: Diagnostic Studies of Low-Density Arc-Heated Wind Tunnel Stream. University of California (No Date).

152. STEM, M. O., and E. N. DACUS: Piezoelectric Probe for Plasma Research. Rev. Sci. Inst., **32**, No. 2 (1961).

153. LI, Y. T.: Dynamic Pressure Measuring Systems for Jet Propulsion Research. J. Amer. Rocket Soc., **23**, No. 3 (1953).

154. The VIDYA Staff.: A Survey of Plasma Diagnostic Techniques. ARL 64-80, U.S. Air Force, 1964.

155. CHEN, C. J.: Velocity Survey in Arc-Jet Tunnel Using Micron Size Tracer Particles. Santa Ana, Calif.: Plasmadyne Corp. (No Date).

156. MCNALLY, J. R., JR.: Role of Spectroscopy in Thermonuclear Research. J. Opt. Soc. Amer., **49** (1959).

157. HURLBUT, F. C.: An Electron Beam Density Probe for Measurements in Rarified Gas Flows. WADC TR-47-644, U.S. Air Force 1958.

158. WINKLER, W.: Interferometry in Physical Measurements in Gas Dynamics and Combustion. Princeton, New Jersey: Princeton University Press, 1954.

159. RAGENT, B., and C. NOBLE: X-ray Densitometer. Report No. 71, VIDYA, April, 1962.

160. ALPHER, R. A., and D. R. WHITE: Optical Refractivity of High Temperature Gases, **I**. Effects Resulting from Dissociation of Diatomic Gases. Phys. Fluids, **2**, No. 2 (1959).

161. ROSNER, D. E.: The Theory of Differential Catalytic Probes. Arlington, Va., AFOSR TN-18, U.S. Air Force, 1960.

162. HARTUNIAN, R. A.: Theory of a Probe for Measuring Local Atom Concentrations in Hypersonic Dissociated Flows at Low Densities. DCAS-TDR-62-101, U.S. Government, 1962.

163. GRIEM, R.: Plasma Spectroscopy. Fifth International Conference on Ionization Phenomena in Gases, **II**, Amsterdam: North Holland Publishing Company, 1962.

164. FUHS, A. E.: An Instrument to Measure the Electrical Conductivity of an Arc-Plasma Jet. American Rocket Society Paper 2635-62, November, 1963.

165. OLSON, R. A., and E. C. LARY: Conductivity Probe Measurements in Flames. American Rocket Society Paper 2592-62, October, 1962.

166. SHAPIRO, A. H.: The Dynamics and Thermodynamics of Compressible Fluid Flow, **I**, New York: The Ronald Press Company, 1958.

167. TRUITT, R. W.: Hypersonic Aerodynamics, New York: The Ronald Press Company, 1959.

168. DORRANCE, W. H.: Viscous Hypersonic Flow. New York: McGraw-Hill Book Company, 1962.

169. GOULARD, R.: On Catalytic Recombination Rates in Hypersonic Stagnation Heat Transfer. Jet Propulsion, **28**, No. 11 (1958).

170. GRINDLE, S. L.: Proposed Recommended Practices for Internal Temperature Measurements in Ablative Materials. ASTM Committee E-21, Section 3 Letter Ballot, October, 1966.

171. DOW, M. B.: Comparison of Measurements of Internal Temperature in Ablation Materials by Various Thermocouple Configuration. NASA D-2165, National Aeronautics and Space Administration, 1964.

172. SKLAREW, S.: The Problem of Accurately Measuring Changing Temperature of Non-Metallic Surfaces. Marquardt Corp., January, 1964.

Appendix

Bibliography of Plasma Arc Technology

A bibliography of literature dealing with flame spraying and plasma arc testing has been compiled. *Section 1* of this bibliography is an alphabetical listing of the principal references for flame spraying and *Section 2* is a compilation of the literature for plasma testing. Most U.S. government documents cited in this bibliography can be obtained by writing to: Department A, National Technical Information Service, U.S. Department of Commerce, Springfield, Virginia 22151, U.S.A.

Section I

Plasma Spraying Literature

1. ALLEN, A. C.: Plasma-Jet A New Tool for Ceramics. Ceram. Ind., **82**, (4) 112—114 (1964).

2. ALLSOP, R. T., T. J. PITT, and J. V. HARDY: The Adhesion of Sprayed Molybdenum. Met., **63**, 125—131 (1961).

3. AL'SHITS, I., and A. M. YA, KORYAVIN: Naneseniie Plastmassovykh Pakriti Na Krupnogabaritnye Izdeliya (Application of Plastic Coatings to Large Articles). Vestn. Mashinostr. **1**, 48—51 (No Date).

4. AMMANN, C., and R. E. CLEARY: Testing of High Emittance Coatings. NASA-CR-1413, National Aeronautics and Space Administration, 1969.

5. ANDERSEN, N.: Development of Plasma Sprayed Gradated Material Systems for Rocket Nozzles and Other Applications. Plasmakote Corp., Contract AF 33 (616)-7323, U.S. Air Force, 1962.

6. ANDERSEN, N.: Arc Sprayed Graded Metal-Ceramic Coatings for Critical High-Temperature Environments. Wright-Patterson Air Force Base, ASD-TDR 63-718, U.S. Air Force, 1965.

7. ANDERSON, J. E., and L. K. CASE: An Analytical Approach to Plasma Torch Chemistry. Ind. Eng. Chem. Process Design Develop., **64**, (1) 161—165 (1962).

8. ANDERSON, L. J.: Strain Gage Applications Using the Rokide Process. Dow Chemical Co., presented at the 9th meeting of IMOG Subgroup on Environmental Testing, Amarillo, Texas, 1965.

9. ANDREEV, V. V., G. V. BOBROV, and M. A. POLEZHAEV: Izgotovlenie Izdelii Iz Molibdena Metodom Plazmennogo Napyleniia (Production of Molybdenum Machine Parts by a Plasma Dusting Process). Poroshkovia Metallurgiia, **5**, (10) 38—46 (1965).

10. ANONYMOUS: Application Data Bulletins. Westbury Long Island, New York: Metco Inc. (No Date).

11. ANONYMOUS: Arc-Sprayed Gradated Coatings. Plasmakote Corp. Contract No. AF 33 (616)-7323, U.S. Air Force, 1961.

12. ANONYMOUS: Ceramics for Jet Use Put into Production. Aviation Week, Space Tech **56**, (6) 30 (1952).

13. ANONYMOUS: Ceramic Systems for Missile Structural Applications. Georgia Institute of Technology Contract No. NOW 63 0143, U.S. Navy, 1963.

14. ANONYMOUS: Ceramic Information Meeting Held at Oak Ridge National Lab. Nucl. Sci. Abstr., **11**, (11) 723 (1957).

15. ANONYMOUS: Coating Articles with Alumina By Flame-Spraying. Norton Grinding Wheel Co., Ltd. Brit., 852, 484 (1960).

16. ANONYMOUS: Critical Compilation of Ceramic Forming Methods. MT 63 12 TDR 63 4069, Institute of Engineering Research University of California (1963).

17. ANONYMOUS: De-Icing with Spraying Metal. Electroplat. Met. Spray, (4) 159 (1954).

18. ANONYMOUS: 30,000 Degrees with the Plasma Jet. J. Metals, **11**, (1) 40—42 (1959).

19. ANONYMOUS: Development of Frontal Section of a Super-Orbital, Lifting Re-Entry Vehicle. Quarterly Progress Report No. 3 Solar Aircraft Co., 1962.

20. ANONYMOUS: Development of a Ferrite Material for a High-Power Phase Shifter at S-Band. Airtron, Inc., Defense Documentation Center AD-443 657, 1964.

21. ANONYMOUS: Development of a Semiconductor Film-Type Thermocouple Energy Converter. Honeywell Research Center, Defense Documentation Center AD-258 075, 1961.

22. ANONYMOUS: Development of High Strength Materials for Solid Rocket Motors. General Electric Co., Defense Documentation Center AD-241 512, 1960.

23. ANONYMOUS: Development Program for Determination of Suitable Materials and Techniques for Fabricating Special Waveguide Assemblies. GB Electronic Corp., Defense Documentation Center AD-269 851, 1961.

24. ANONYMOUS: Detonation Gun Explodes Old Flame-Spraying Ideas. Prod. Eng., **37**, (17) 62—65 (1966).

25. ANONYMOUS: The Effect of Arc Plasma Deposition on Stability of Non-Metallic Materials. General Electric Co., Flight Propulsion Division, Contract Noas-60-6976-C, U.S. Navy, 1960.

26. ANONYMOUS: Evaluation Methods and Equipment for Flame Sprayed Coatings. Report No. 106, Long Island, New York: Metco Inc. (No Date).

27. ANONYMOUS: Fabrication of Pyrolytic Graphite Solid Rocket Nozzle Components. General Electric Co., Defense Documentation Center AD-243 906, 1960.

28. ANONYMOUS: Finishes for Metal Products. Mater. Methods Manual No. 119, September 1955.

29. ANONYMOUS: Flame Plating Process. Steel, **135**, (8) 67 (1954).

30. ANONYMOUS: Flame-Plating Applicable to Many Metals and Alloys. Amer. Metal Market, (6) 4 A (1966).
31. ANONYMOUS: Flame Plating Makes Tools Last Longer. Welding Design Fabric., **35**, (12) 67 (1962).
32. ANONYMOUS: Flame Spraying Multiplies Life of Worn Parts for Block Making Machinery. Metco Report, Long Island, New York: Metco Inc. (No Date).
33. ANONYMOUS: The Effect of Plasma Spray Coatings on Fatigue Properties of H-11 Steel and 2024 Aluminum. Metallurgical Report B-714, Curtiss-Wright Corp. Wood-Ridge N.J., March, 1971.
34. ANONYMOUS: Flame Sprayed Ceramic Coating Techniques, Economic Summary. Reaction Motors Division Thiokol, Defense Documentation Center AD-464-847, 1964.
35. ANONYMOUS: High-Frequency Metal Spraying. Science and Technology Division Library of Congress, Defense Documentation Center AD-266 483 (No Date).
36. ANONYMOUS: High-Temperature Oxidation Prevented by Sprayed Aluminum. Aluminum Electroplat, (2) 74 (1953).
37. ANONYMOUS: High-Temperature Oxidation Protective Coatings for Vanadium-Base Alloys. IIT Research Institute Contract NOw 61-0806-C, U.S. Navy, 1962.
38. ANONYMOUS: High-Temperature Oxidation Protective Coatings for Vanadium-Base Alloys. IIT Research Institute, Defense Documentation Center, AD-417 137, 1963.
39. ANONYMOUS: The Importance of the Plasma Jet. Machinery, **98**, (4) 807, 858 (1961).
40. ANONYMOUS: Inorganic Dielectric Research. New Jersey Ceramic Research Station, Rutgers University, Defense Documentation Center AD-285 691, 1962.
41. ANONYMOUS: The Aerospace Manufacturing Techniques Panel of the Committee on the Development of Manufacturing Processes for Aircraft Materials. Materials Advisory Board, Defense Documentation Center AD-296 324, 1963.
42. ANONYMOUS: Report of the Ad Hoc Committee on Processing of Ceramic Materials. Materials Advisory Board, Defense Documentation Center, AD-421-422, 1963.
43. ANONYMOUS: The Metallizing Process. Welding J., **31**, (4) 291—295 (1952).
44. ANONYMOUS: Module Improvement Program. Westinghouse Electric Corp., Defense Documentation Center AD-282 505, AD-273-657, 1962.
45. ANONYMOUS: Metallized Transmission Parts Last Longer. Metco Report, Long Island, New York: Metco Inc. (No Date).
46. ANONYMOUS: Molybdenum Alloy Extrusion Development Program. Allegheny Ludlum Steel Corp., Wright-Patterson Air Force Base ASD TR 7-7850 U.S. Air Force, 1962.
47. ANONYMOUS: Mounting Sensing Elements with Rokide Ceramic Coatings. Norton Co., 1963.
48. ANONYMOUS: Metal Surfacing for Original Parts. Prod. Eng., **22**, (8) 124 (1951).
49. ANONYMOUS: Metallizing Bibliography. American Welding Society, New York, 1956.
50. ANONYMOUS: Metallizing of Plastics. Metall., **8**, (17—18) 675 (1954).
51. ANONYMOUS: Method for Measuring Bond Strength of Sprayed Metal Coatings. Metco Report, Long Island, New York: Metco Inc., 1957.
52. ANONYMOUS: New Electric Forming Methods Offer Potentials for Light Metals; Magnetic Forming and Plasma Forming. Light Metal Age, **20**, (2) 8—9 (1962).
53. ANONYMOUS: The Plasma Arc-A New Tool for Welding and Cutting. Welding Fabrication and Design, **3**, (12) 10—11 (1960).
54. ANONYMOUS: Plasma Arc Metal Spraying. Metal Ind., **101**, (1) 11—12 (1962).
55. ANONYMOUS: Plasma Arc-Multi-Capability for Economy. Welding Eng., **43**, (4) 57—58 (1958).
56. ANONYMOUS: Plasma Arc Process Coats and Forms Refractory Material. Space Aeron., **32**, (6) 105—109 (1959).
57. ANONYMOUS: Plasma Arcs Set for Full Production Role. Steel, **148**, (5) 70—72 (1961).

58. ANONYMOUS: The Plasma Arc Torch. Welding and Metal Fabric., **27**, (7) 287—290 (1959).

59. ANONYMOUS: Plasma Arc Torch Applies Unworkable Materials. Machinery, **65**, (5) 122 (1959).

60. ANONYMOUS: The Plasma-Arc Torch Finds a Wide Range of Tasks. Welding J., **39**, (3) 236—237 (1960).

61. ANONYMOUS: The Plasma Arc Torch Users in New Fabricating. Coating Methods. Iron Age, **182**, (12) 136—137 (1959).

62. ANONYMOUS: Plasma Coating; Tailored Solution for More Jobs. Steel, **156**, (21) 80—83 (1956).

63. ANONYMOUS: Plasma Flame Spraying Equipment. Eng., **213**, (6) 1093 (1962).

64. ANONYMOUS: Jet de Plasma (Plasma Jet). Le Moniteur Professional de L'Electrocite, **16**, (166) 32 (1961).

65. ANONYMOUS: Plasma Torches Cut Metals with Arc-Heated Air: Save Time and Expense. Marine Eng. Log., **67**, 103 (1962).

66. ANONYMOUS: Plasma Torch Learns to Weld. Amer. Machinist Metalworking Manu., **105**, (4) 115 (1961).

67. ANONYMOUS: La torche à plasma est Maintenance Industrielle (Present Industrial Uses of the Plasma Torch). L'Industrie Française, **10**, (113) 700—710 (1961).

68. ANONYMOUS: Present Uses, Future Hopes for Plasma E.B. Laser. Welding Eng., **49**, (1) 35 (1964).

69. ANONYMOUS: Prevention of Corrosion on Loading Rack Tee Tracks and Launching Rail Wheels, Model-Nike-Hercules. Bell Telephone Labs. Inc., Defense Documentation Center AD-244 168, 1960.

70. ANONYMOUS: Properties of Plasma Sprayed Materials. General Electric Co., Contract AF 33 (616)-6376, U.S. Air Force, 1960.

71. ANONYMOUS: Properties of Flame Sprayed Metal and Metal Oxide Coatings. Boeing Airplane Company (No Date).

72. ANONYMOUS: Protective Coatings for Iron and Steel Compared. Electroplat. Met. Spray., (5) 207, 1954.

73. ANONYMOUS: Protective Coatings for Refractory Metals in Rocket Engines. IIT Research Institute, Report No. IITRI-B 0 237-12, 1963.

74. ANONYMOUS: Report No. 2 of the Aerospace Manufacturing Techniques Panel of the Committee on the Development of Manufacturing Processes for Aircraft Materials. Iowa State University, Defense Documentation Center AD-462 323, October, 1963.

75. ANONYMOUS: Shop Notes for Material Spray Applications with Plasmadyne S-Series Hand Spray Guns. Plasmadyne Corp. (No Date).

76. ANONYMOUS: Spray Coatings. Welding Eng., **50**, (7) 37—46 (1965).

77. ANONYMOUS: Soviet Ceramics and Refractories; (Selected) Articles. Joint Publications Research Service, Defense Documentation Center—380, AD 298-830, 1962.

78. ANONYMOUS: Semi-Annual Review of Material Work in Progress. Aero-jet-General Corp., Defense Documentation Center, AD-4X12 400, 1963.

79. ANONYMOUS: Relationships Between Cavitation Erosion and Fundamental Properties of Materials. Naval Applied Science Lab., N.Y., Defense Documentation Center, AD-255 611 L, 1961.

80. ANONYMOUS: Translation in Communist China's Science and Technology, No. 301. Joint Publication for Research Service Washington, D. C., 34620, 1966.

81. ANONYMOUS: Symposium on Porcelain Enamels and Ceramic Coating on Engineering Materials. American Society for Testing Materials, 1954.

82. ANONYMOUS: Technology's Newest Baby: The Plasma Torch. Machine Moderne, **53**, (5) 9—12 (1959).

83. ANONYMOUS: Thermal Studies of 81-mm Titanium Mortar Barrel. Aberdeen Proving Ground, Defense Department Center, AD-268 996, 1961.
84. ANONYMOUS: Torch Cuts with 60,000°-F Air. Mech. Eng., **84**, (7) 66 (1962).
85. ANONYMOUS: Thermionic Double-Diode Programm. Martin Co., TDR 63 4244, November, 1963.
86. ANONYMOUS: Tungsten Carbide Flame Plating Investigation. Aerojet-General Corp., Defense Documentation Center AD-427 504, 1963.
87. ANONYMOUS: Where Coatings not Wanted, Salt Bath Removes It. Steel, **156**, (13) 126—128 (1965).
88. ANONYMOUS: Where is Plasma Arc Today? Amer. Machinist, **103**, (22) 125—127 (1959).
89. ANTOINE, R.: Mise En Œuvre Et Etude Des Jauges D'Extensométrie Fixées Par Le Procédé ROKIDE. (Use and Study of Strain-Gauges Attached by the ROKIDE Process). European Atomic Energy Community, ISPRA, Italy, 1969.
90. APPEN, A. A.: Heat Resistant Inorganic Coatings. Wright-Patterson Air Force Base, FTD-MT-24-497-68, U.S. Air Force, 1969.
91. ARENBERG, C. A.: Protective Coatings for Tantalum. Wright-Patterson Air Force Base, WADC TR-58-203, U.S. Air Force, 1958.
92. ARNOLD, H.: Das Metallspritzverfahren, seine wissenschaftlichen, technischen und wirtschaftlichen Grundlagen (Metal Spraying Technology Its Scientific, Technical and Economic Principles). Z. F. angewandte Chemie, **30**, 209—214, 218—220 (1970).
93. ASKWYTH, R. J., HAYES, and G. MIKK: The Emittance of Materials Suitable for Use as Spacecraft Radiator Coatings. American Rocket Soc., Space Power Systems Conference Paper 2538—2562, 1962.
94. AST, F.: Bestimmung der beim Metallspritzprozess auftretenden Temperaturfelder und technologische Untersuchungen an metallgespritzten Schichten (Determination of the Temperature Fields Arising from the Metal Spraying Process and Technological Investigations of Metal Sprayed Layers). Diss. T. H., Munich, 1954.
95. AULT, N. N.: Characteristics of Refractory Oxide Coatings Produced by Flame Spraying. J. Amer. Ceram. Soc., **40**, (3) 69—74 (1957).
96. AULT, N. N., and L. H. MILLIGAN: Alumina Radomes by Flame-Spray Process. Amer. Ceram. Soc. Bull., **38**, (11) 661—664 (1959).
97. AULT, N. N., and W. M. WHEILDON: Modern Flame Sprayed Ceramic Coatings. Modern Materials, **2**, ed. HAUSNER, H. H., New York: Academic Press (1960).
98. AVES, W. L.: Coatings for Re-entry. Metal Prog., **75**, (3) 90—94, 189C, 190 (1959).
99. BABBITT, R. W., and W. W. MALINOFSKY: Control and Reproducibility of Ferrites Prepared by the Flame Spray-Hot Press Technique. Army Electronics Command A 0070657, U.S. Army, 1965.
100. BACON, J. F., and R. D. VELTRI: Improved Method of Electrical Contact to Ceramics for High-Temperature Uses. Rev. Sci. Instr., **34**, 1264—1265 (1963).
101. BAKER, C., and S. J. GRISAFFE: Analysis of the Shear-Bond Strength of Alumina Coatings. NASA-TM-X-56900, National Aeronautics and Space Administration, 1965.
102. BALLARD, W. E.: The Formation of Metal-Sprayed Deposits. Proc. Phys. Soc., **57**, Part 2, (320) 67—83 (1945).
103. BALLARD, W. E.: Metal Spraying and Sprayed Metal. London, C. Griffin and Co., Ltd., 1948.
104. BARDWELL, R. A.: Flame Ceramics. Mod. Railr., (4) 1959.
105. BARLAND, E. S., S. R. ELKINS, F. L. JONES, and B. WILKINS JR.,: Arc Plasma Coatings in Refractory Metals. Refractory Metals and Alloys III, Applied Aspects, New York: Gordon and Breach Science Publishing Inc., 1966.
106. BARNETT, C. W. H.: Ultrathermic Capacitor. Wright-Patterson Air Force Base, WADD TR-61-50, U.S. Air Force, 1961.

107. BARNETT, C. W. H.: 500 C Capacitors (Refractory Dielectrics). Defense Documentation Center, AD-231 309, 1959.
108. BARTA, I. M.: Low Temperature Diffusion Bonding of Aluminum Alloys. Welding J. Res. Supplement, **43**, (6) 241-S—247-S (1964).
109. BARTH, V. D.: Review of Recent Developments in the Technology of Tungsten. Defense Metals Information Center, 1961.
110. BARTH, V. D.: Review of Recent Developments-Powder Metallurgy. Defense Metals Information Center, 1963.
111. BARTH, V. D.: Review of Recent Developments, Tungsten and Tungsten-Base Alloys. Defense Metals Information Center, 1965.
112. BATCHELOR, L. E.: Sprayed Aluminum Reduces Compressor Clearance. Amer. Machinist, **102**, (10) 140—141 (1958).
113. BAZZARRE, D., and L. J. FRANKLIN: Improvement of Interlaminar Shear Strength in Organic Matrix Composites Utilizing Plasma Deposited Coupling Agents. Quarterly Report 2, Contract #NOOO 17-69-C 4429, U.S. Navy, 1969.
114. BELANGER, J. Y., and F. CHRISTIE: The Protection of Carde Rocket Nozzles by Flame Sprayed Ceramic Coatings. Carde-TN-1684, Canadian Armament Research Development Establishment, 1965.
115. BENNETT, D. G.: Suppression of Radiation of High-Temperatures by Means of Ceramic Coatings. J. Amer. Ceram. Soc., **30**, (10) 297 (1947).
116. BENNETT, D. G., et al.: High-Temperature Resistant Ceramic Coatings, Ceramic and Metal Ceramic Bodies. AD 38003 Defense Documentation Center, 1955.
117. BERG, R. E.: Plasma Cutting Torches. Amer. Machinist, **105**, (24) 93—96 (1961).
118. BERTOSSA, R. C.: New Metals and Processes. Metal Treat., **9**, (6) 2—4 (1958).
119. BITONDO, D., N. THOMAS, and D. PERPER: Non-Stationary Surface Combustion Studies. Defense Documentation Center, AD-139 186 (No date).
120. BLAMPIN, B.: Lubrification Par Fluorure de Calcium dans L'Air et Le Gaz Carbonique Chauds (Lubrication by Calcium Fluoride in Hot Air and Carbon Dioxide). Wear, **11**, (7) 431—459 (1968).
121. BLANCHARD, J. R.: Oxidation Resistant Coatings for Molybdenum. Wright-Patterson Air Force Base, WADC TR-54-492, U.S. Air Force, 1954, 1955.
122. BLITON, J. L., W. J. CHRISTIAN, J. W. DALLY, J. C. HEDGE, and H. H. HIRSCHHORN: Evaluation of Thermal Protective Systems for Advanced Aerospace Vehicles. Wright-Patterson Air Force Base, ML-TDR-64-204, **1**, U.S. Air Force (1965).
123. BLITON, J. L., and R. HAVELL: Physical Properties of Flame-Sprayed Ceramic Coatings. Part II, Barium Titanate, Amer. Ceram. Soc. Bull., **41**, (11) 762—767 (1962).
124. BLITON, J. L., and J. J. RAUSCH: Plasma-Sprayed Oxide and Vapor-Deposited Nitride Coatings on Tungsten as a Means of Achieving Oxidation Protection. NASA Report No. NAS 7-113, National Aeronautics National and Space Administration, 1963.
125. BLITON, J. L., H. L. RECHTER, and Y. HARADA: Flame Sprayed Zirconia Films for Fuel Cell Components. Amer. Ceram. Soc. Bull., **42**, (1) 6—9 (1963).
126. BLITON, J. L., and H. L. RECHTER: Determination of Physical Properties of Flame-Sprayed Ceramic Coatings. Amer. Ceram. Soc. Bull., **40**, (11) 683—688 (1961).
127. BLITON, J. L., and H. L. RECHTER: The Design of a Flame-Spray Coating. Solid Bodies, Society of Aerospace Materials and Process Engineers, North Hollywood, Western Periodicals Company, 1961.
128. BLITON, J. L., and S. A. BORTZ: Ceramic Coatings for Cementitious and Metallic Surfaces. Illinois Institute of Technology IITRI-B 8009-7, 1964.
129. BLOOM, D. S.: Tungsten and Rocket Motors. Defense Documentation Center, AD-263 118 (No Date).

130. Blumberg, L. N., et al.: Electrospraying of Thin Targets. Los Alamos Scientific Lab., N 62-15631, 1962.

131. Bober, E. S., W. H. Snavely, and R. E. Stapleton: Development of High-Temperature Alkali Metal Resistant Insulated Wire. Contract AF 33657 10701, U.S. Air Force, 1964.

132. Bollinger, L. E., and R. Edse: Formation of Detonation Waves in Flowing Combustible Gaseous Mixtures and Liquid-Spray Mixtures. Ohio State University, Defense Documentation Center, AD-631, 786, 1965.

133. Bradley, E. F.: Hard Facing Upgrades Turbine Engine Components. Metal Progr., **88**, (11) 79—82 (1965).

134. Bradley, E. L., and S. D. Stoddard: An Arc Spray Powder Feeder for Ultra-Fine Particles with a Wide Range of Densities. Presented at the Pacific Coast American Ceramic Society Meeting, 1965.

135. Bradstreet, S. W.: Ceramic Coatings Present and Future. Presented at the 63rd Annual Meeting of the American Ceramic Society, 1961.

136. Bradstreet, S. W.: Flame Sprayed Catalyst Coatings. Presented at the 22nd Shop Practice Forum of the Professional Engineers Institute, 1960.

137. Bradstreet, S. W.: Flame Ceramics. Frontier, **19**, (4) 4 (1956).

138. Brandes, R. G.: Plasma Deposition. Bell Lab. Record, **43**, (4) 288—292 (1965).

139. Brinkman, R. J., ed.: Ceramic Coating Conference, 27 and 28 May, 1952. Section I, Adherence Tests, Wright-Patterson Air Force Base, WADC TR-53-37, U.S. Air Force, 1954.

140. Brophy, J. H., H. W. Hayden, K. G. Krieder, and J. Wulff: Activated Sintering of Pressed Tungsten Powders and Plasma-Jet-Sprayed Tungsten Deposits. Bureau of Naval Weapons, Access No. 12, 467, U.S. Navy, 1961.

141. Brophy, J. H., J. R. Heideklang, K. F. Krieder, and J. Wulff: Activated Sintering of Pressed Tungsten Powders and Plasma-Jet-Sprayed Tungsten Deposits. Contract NOa (s)-59-6264-c, U.S. Navy, 1959—1960.

142. Brouillette, C. V., and R. L. Alumbaugh: Cost Comparison of Protective Coatings for Steel Piling. Naval Civil Engineering Lab. TR-490, Suppl., U.S. Navy, 1967.

143. Brown, S. D.: Room-Temperature Adhesive Tests of Various Flame-Sprayed and Flame-Plated Ceramic Coating. Ordnance Corps., Prog. Rept. No. 20-374, U.S. Army, 1959.

144. Browning, J. A.: Thermal Air Cutting: Process and Economics. Welding J., **41**, (6) 458—466 (1962).

145. Browning, J. A.: Machining With a Plasma Jet. Amer. Machinist/Metalworking Manu., **106**, (6) 94—95 (1962).

146. Browning, J. A.: Plasma Flame Speeds Metalworking. Tool Eng. **44**, (4) 105—108 (1960).

147. Browning, J. A.: Plasma Arc Speeds Many Metalworking Processes. Amer. Inst. Tool Manu. Eng., **60**, Book 1, Tech. Paper 288 (No Date).

148. Browning, J. A.: Techniques for Producing Plasma Jets. Gas Dynamics Symposium, 1959.

149. Browning, J. A.: Plasma—A Substitute for the Oxy-Fuel Flame. Welding J., **38**, (6) 870 (1959).

150. Browning, M. E., H. W. Leavenworth, W. H. Webster, and F. J. Dunkerly: Deposition Forming Processes for Aerospace Structures. Wright-Patterson Air Force Base, ML TDR 64-26, 1959.

151. Brzozowski, W., M. Mikos, and J. Reda: Vnedrenie Plazmennoi Tekhniki V Khimicheskuiu I Metallurgicheskuiu Promyshlennost (Introduction of Plasma Technology into the Chemical and Metallurgical Industry). Transactions Nizkotemperaturnaia Plazma,

Mezhdunarodnyi Kongress Po Teoreticheskoi I Prikladnoi Khimii, 4720th Mezhdunarodyni Simpoziumpo Svoistvam I Primeneniiu Nizkotemperaturnoi Plazmy, Moscow, USSR, July, 1965.

152. BUCKLEY, J. D.: Thermal Conductivity and Thermal Shock Qualities of Zirconia Coatings on Thin Gage Ni-Mo-C Metal. Amer. Ceram. Soc. Bull., **49**, (6) 588—591 (1970).

153. BUFFINGTON, J. W.: Metal Spraying with Rockets. Welding J., **35**, (5) 468 (1956).

154. BUGINAS, S. J.: Plasma Jet Welding, Coating and Cutting; an Annotated Bibliography. Lockheed Missiles and Space Co., 1963.

155. BURKE, M. H.: Future Development in Plasma Arc Welding. Aircraft Eng., **40**, (12) 13—15 (1968).

156. BURKE, M. H.: Symposium on Welding in the Aerospace Industry. Society of British Aerospace Companies, 1968.

157. BURROUGHS, J. E., J. E. HALKIAS, and H. R. THORNTON: Plasma Arc Deposition for Aerospace Applications. Presented at the 3rd Aerospace Finishing Symposium, American Electroplaters Society, January, 1964.

158. BURYKINA, A. L., and T. M. YEVTUSHOK: Plazmennye I Diffuzionnye Pokrytiia Na Grafite (Plasma Coatings and Diffusion Coatings on Graphite). Poroshkovaia Met Nauk Ukr SSR, **5**, 39—44 (1965).

159. BUTTA, M.: A New Type of Cutting Torch: The Plasma-Torch. Machine, **8**, 791 (1959).

160. CAUCHETIER, J.: Sprayed High Melting Point Metals. Metal Ind., **93**, (18) 374—375 (1958).

161. CAUCHETIER, J.: Metallization by Wire Pistol with Light Alloys. Rev. Aluminum, **179**, 297 (1951).

162. CAUCHETIER, J.: The Adherence Obtained When Spraying Metals with Very High Melting Points. Given at the Second International Metal Spraying Symposium, Birmingham, 1958.

163. CAUGHEY, R. A., F. M. WILSON, and R. A. LONG: Alumina Radome Attachment. Defense Documentation Center, AD-271 588, 1962.

164. CHAPIN, E. J., and W. A. REAVES: An Investigation of Barrier Coatings on Graphite Molds for Casting Titanium. Naval Research Laboratory, NRL-6295, U.S. Navy, 1965.

165. CHEREPANOV, A. M.: High-Temperature Inorganic Coatings for Modern Technology. Zh. Vses. Khim. Obshchestva im. D. I. Mendeleeva, **10**, (5) 532—539 (1965).

166. CHEVELA, O. B., and L. M. ORLOVA: O Formirovanii Plazmennykh Pokrytii Na Poverkhnosti Stalei I Titanovykh Splavov (Formation of Plasma Coating on the Surface of Steels and Titanium Alloys). Poroshkvaia Met Nauk Ukr SSR, **8**, (12) 26—34 (1968).

167. CHILD, M. R., and A. L. LEVETT: Spraying Equipment for Oxide Cathodes. Defense Documentation Center, AD-442 943 L, 1962.

168. CHRISTIANSEN, F.: Plasma Torch: Applied Plasma Physics. Nag-Tidsskriftet, **25**, (3) 42—44 (1961).

169. CLARK, J. W., and C. H. WODKE: Plasma Arc Torch Cuts Off "Hot" Plates. Iron Age, **188**, (7) 90—91 (1961).

170. CLIFF, A.: Protection of Tubular Structures from Weather by Metallizing. Metal Progr., **61**, (5) 100 (1952).

171. CLOSE, C. G.: Metallizing for Corrosion Prevention. Proc. Eng., **23**, (7) 58 (1952).

172. CLOUGH, P. J., and P. L. RAYMOND: Oxidation Resistant Coatings for Molybdenum and Other Refractory Bodies. Amer. Ceram. Soc. Bull., **40**, (5) 314—315 (1961).

173. COOLEY, R. A., C. M. HENDERSON, R. J. JANOWIECKI, and M. C. WILLSON: A New Technique for Fabricating Thermoelectric Elements and Generators. Institute of Electrical and Electronics Engineers and American Institute of Aeronautics and Astronautics, Thermoelectric Specialists Conference, Washington, May, 1966.

174. COPELAND, R. L., and V. A. CHASE: Supersonic Rain Erosion Resistant Coating Materials Investigation. Wright-Patterson Air Force Base, TR-67-62, U.S. Air Force, 1967.

175. COWEN, J. M.: Adhesion and Surface Preparation in Protective Metal Spraying. Electroplat. Met. Spray., (2) 79—82 (1954), (3) 117—122 (1954).

176. CREMER, F.: Das Metallspritzverfahren (Metal Spraying Technology). Der Eisenbahningenieur, 10, (12) 362—366 (1959).

177. CROSBY, A., and P. SYKES: The Manufacture of Ceramic Radomes by the Flame Spray Process. Délégation Ministérielle Pour L'Armement, International Symposium on Electromagnetic Windows University of Paris, France, 1967.

178. CROUCHER, T. R.: Wear Resistance of Sprayed Metallic and Ceramic Coatings. Defense Documentation Center AD-445 190, 1963.

179. CROWLEY, J., LT.: Literature Review of the Application of Sprayed Metals to Metal and Nonmetal Surfaces. Wright-Patterson Air Force Base, WCRT TM 56-115, U.S. Air Force, 1956.

180. DALLMAN, A. C.: Mechanical Reliability and Thermoelectrical Stability of Noble-Metal Thermocouples at 2600° F Temperature and Dose Rates up to 10 to 20 Power NVT. Wright-Patterson Air Force Base, TDR 64-7, U.S. Air Force, 1964.

181. DAMON, R. A., R. L. LANDINGHAM, and D. J. MURPHY: Plasma-Jet Coating of Tungsten on Steel. The Fusion Bonding of Plasma-Jet Coatings to Metals. Defense Documentation Center AD-625 800, 1962.

182. DAVIS, L. W.: How to Deposit Metallic and Nonmetallic Coatings with a Plasma Arc Torch. Metal Progr., 83, (3) 105—108 (1963).

183. DAVIS, L. W.: How Metal Matrix Composites are Made. Amer. Soc. for Testing and Mater., No. 427, 69—90, 1967.

184. DENNIS, P. R., C. R. SMITH, D. W. GATES, and J. B. BOND: Plasma Jet Technology. NASA SP-5033, National Aeronautics and Space Administration, 1965.

185. DEPAAUH, D.: Porosity of Sprayed Coating. Parts 1—4, Electroplat. Met. Spray., (11) 435—438, (12) 475—476 (1953), (1) 36—38, (1954).

186. DICKINSON, T. A.: Flame Spray Ceramics. Prod. Finish., 33, (2) 32 (1954).

187. DICKINSON, T. A.: Flame Spraying Ceramic Coatings. Prod. Finish., 41, (3) 84 (1958).

188. DIETZEL, A.: Praktische Bedeutung und Berechnung der Oberflächenspannung von Gläsern, Glasuren und Emails. Sprechsall, Keram., Glas, Email, 75, 82—85 (1947).

189. DITTRICH, F. J.: New Flame Spray Techniques for Forming Nickel Aluminide-Ceramic Systems. Amer. Ceram. Soc. Bull., 44, (6) 492—496 (1965).

190. DITTRICH, F. J.: Thermal Shock Resistance of Plasma Sprayed Zirconia Coatings. Presented at the 72nd Annual Meeting of the American Ceramics Society, 1970.

191. DOANE, D.: Oxidation Resistant Coatings for Molybdenum. Wright-Patterson Air Force Base, WADC TR 54-492, Part III, U.S. Air Force, 1957.

192. DOLIWA, H.: Gespritzte Metallüberzüge (Sprayed Metal Coatings). Ind.-Anz., 81, (84) 1331—1333 (1959).

193. DONNELLY, W. I.: Carbide Flame-Plating. Amer. Soc. Tool Eng., Technical Papers and Panel Discussions, 21, (1953).

194. DONOVAN, M.: Experience in the Use of Plasma Spraying Techniques. Brit. Welding J., 13, (8) 490—496 (1966).

195. DORSEY, J.: Development and Evaluation Services on Ceramic Materials and Wall Composites for High-Temperature Radome Shapes. Wright-Patterson Air Force Base, WADC TR 57-665, ASTIA # 15-965, U.S. Air Force, 1958.

196. EARLE, F. M.: Metallizing Cuts Marine Maintenance Cost. Iron Age, 169, (5) 103 (1952).

197. ECKERSLEY, A.: H-Field Shielding Effectiveness of Flame-Sprayed and Thin Solid Aluminum and Copper Sheets. IEEE Transactions on Electromagnetic Compatibility (EMC-10) 101—104, 1968.

198. EISENLOHR, A.: Properties of Plasma-Sprayed Materials. Contract No. 33 (616)-6376, U.S. Air Force, 1960—1961.

199. EISENLOHR, A., and H. SPECHT: Arc Plasma Spray Gun Design for Wire Deposition. Internal G. E. Report DM 59-341, General Electric Co. (No Date).

200. ELKINS, D. A., and C. H. SCHAEK: Possible Applications of Plasma Technology in Minerals Processing. TN 23. U 71, No. 8438, Bureau of Mines, U.S. Dept. of the Interior, 1969.

201. ELSTON, D. L., and G. BRODI: Plasma-Sprayed Refractory Coatings on Small-Scale Graphite Crucibles. National Lead Co. of Ohio, N 62-17516, 1962.

202. ELSTON, D. L.: The Plasma Spraying of Protective Coatings. Ann. Natl. Lead Co. Anal. and Phys. Testing Symp., New Jersey, 1965.

203. ENGELKE, J. L.: Heat Transfer to Particles in the Plasma Flame. Presented at the A.I.Ch. E. Meeting, Los Angeles, Calif., February, 1962.

204. ENGELKE, J. L., F. A. HALDEN, and E. P. FARLEY: Synthesis of New High-Temperature Materials. Wright-Patterson Air Force Base, WADC TR 59-654, U.S. Air Force, 1958—1959.

205. ENGELL, H.: Die Haftfestigkeit von Oberflächenschichten auf Metallen (The Adhesive Strength of Surface Layers on Metals). Werkstoff Korrosion, 11, (3) 147—151 (1960).

206. EPPINGER, E. D., T. A. GREENING, and S. M. JACOBS: Wire Wound Plasma Spray Bonded Tungsten Solid Rocket Nozzle Insert Materials. New York, American Inst. of Aeronautics and Astronautics, 1968.

207. EORGAN, J. E., and N. D. FERN: Zirconium Diboride Coating on Tantalum. J. Metal, 19, (9) 6—11 (1967).

208. ERICKSON, G. L.: Properties of Flame Sprayed Metal and Metal Oxide Coatings. Boeing Airplane Co., 1959.

209. ESCHENBACH, R. C., and R. J. WICKHAM: Metal-Ceramic Ablative Coating. Wright-Patterson Air Force Base, WADD TR 60-439, U.S. Air Force, 1959—1960.

210. ESHELMAN, R. H.: Plating with Carbide. Tool Eng., 36, (1) 117—122 (1956).

211. ESTY, C. C., R. W. LOVE, and W. M. WHEILDON: High-Temperature Materials and Flame Spray Coatings for the Aerospace Industry—Their Processing, Characteristics, and Applications. Material Science Research Proceedings of the 1964 Southern Metals-Materials Conference on Advances in Aerospace Material, Vol. 2, Edited by Otte and Locke, New York: Plenum Press, 1965.

212. ESTY, C. C., W. H. MCMAKEN, and W. M. WHEILDON: Corrosion Protection Via Flame Spraying. AFML 50th Anniv. Symp. on Corrosion of Military and Aerospace Equipment, Denver, U.S. Air Force, 1967.

213. FABIAN, R. J.: Hard Coatings and Surfaces for Metals. Mater. Methods, Manual 134 1956.

214. FABIAN, R. J.: New Coatings from the Plasma Arc. Mater. Design Eng., 54, (6) 127—138 (1961).

215. FABIAN, R. J.: Plated Cermet Coatings Fight Heat Wear. Mater. Design Eng., 56, (3) 105 (1962).

216. FABIAN, R. J.: What's New in Coatings and Finishes. Mater. Design Eng., 56, (4) 109 to 116 (1963).

217. FABIAN, R. J.: Wear Resistant Materials and Coatings. Mater. Design Eng., 56, (6) 131—146 (1962).

218. FAIRLIE, J.: Plasma For Cutting, Welding, Coating. Welding Eng., 47, (11) 41—44 (1962).

219. FARROW, R. L., and M. LEVY: Analysis of a Refractory Coating System for the Thermal Protection on Titanium. Defense Documentation Center AD-421-816, 1963.

220. FINCH, N. J., and J. E. BOWERS: Surface Treatment of Titanium Alloys: A Review of Published Information. British Non-Ferrous Metals Research Association London, England, Defense Documentation Center AD-469 954, 1965.

221. FISCHER, G. W.: Which Metal Spray Coating: Plasma or Metallizing? Machinery, **68**, (8) 83—90 (1962).

222. FISCHER, G. W.: Comparison of Metal Spraying by the Plasma-Arc and Gas Flame Processes. Machinery, **101**, (12) 1423—1428 (1962).

223. FORD, R. D.: Evalation of Powder Dispensers for Use in Flame Spraying. Rubber Lab., Mare Island Naval Shipyard, Vallejo, Calif., Defense Documentation Center, AD-147 701 (No Date).

224. FOX, H. A., JR.: Flame Spray Capability Tests for Titanium, Hastelloy, 347 Stainless Steel, and 6061 T-6 Aluminum. RN-TM-301, Aerojet General Corp., Sacramento, California, 1968.

225. FRANKLIN, J. R.: Metallized Coatings for Heat Corrosion Protection. Corrosion Tech., **3**, (10) 326 (1956).

226. FRIEDMAN, E., R. W. ALLAN, and G. R. WEISSMAN: Development of a Sprayable, Strippable, Protective Coating for Aircraft, Rockets and Missiles. Snell Inc., New York, Defense Documentation Center, AD-287 515 (No Date).

227. FRITZ, J. C.: Flammspritzen von Stahl, Metallen und Kunststoffen. Essen: Verlag W. Giradet, 1955.

228. FROLOV, A. S., M. G. TROFIMOV, and E. M. VERENKOVA: Gazoplamennoe Napylenie Pokrytii Iz ZrO₂ I Al₂O₃ Dobavkoi Aliumofosfata (Gas Flame Spraying of ZrO_2 and Al_2O_3 with Aluminum Phosphate Additives). Transactions Vysokotemperaturnye Pokrytiia, Seminar Po Zharostoikim Pokrytilam Leningrad, USSR, 1965.

229. FULLER, L. E.: Mating Materials in Unlubricated, High Load, Low Speed, Wear Tests at High-Temperature in Air. Soc. Automotive Eng. J., **68**, (12) 60—65 (1960).

230. GAGE, R. M.: The Plasma Arc Can Now Be Used for Welding and Weld Surfacing. Welding Design Fabric., **34**, (4) 76—78, 80 (1960).

231. GAGE, R. M., and J. F. PELTON: Plasma-Arc Torch: Five Years of Evolution. ASME Paper 60-WA-311 (No Date).

232. GALDON, B. F.: Special Study of a Method for Attaching Thermocouples to Mortar Tubes. Defense Documentation Center, AD-488-897 L, 1966.

233. GALLEGER, J.: Protection of Steelwork by Sprayed Aluminum. Electroplat Met. Spray., (1), 1954.

234. GALLI, J., G. WHEELER, B. CLAMPITT, D. GERMAN, and R. JOHNSON: Development and Evaluation of Rocket Blast and Rain Erosion Resistant Composite Coatings Produced by Flame Spray Techniques. Wright-Patterson Air Force Base, WADC TR 58-493, U.S. Air Force, 1959.

235. GAON, L., and J. L. CIRINGIONE: Study of Effects of Metallizing Procedures on the Fatigue Strength of Crankshafts. New York Naval Shipyard, Defense Documentation Center, AD-225 238 L, 1958.

236. GARIBOTTI, D. J.: Modular Interconnections for Micro-Assemblies-Phase II. United Aircraft Corp., Defense Documentation Center, AD-287 942, 1962.

237. GATZEK, L. E.: New Developments in Wear-Resistant Finishes and Coatings. Amer. Soc. of Mechanical Eng., Design Engineering Conference and Show, Paper 63-MD-9, 1963.

238. GEORGI, H.: Möglichkeiten des Spritzschweißprozesses. (Possibilities of Spray Welding Processes). Schweißtech., **7**, (4) 130—137 (1957).

239. GERHOLD, E. A.: Hard-Facing with Plasma Spray Guns. Brit. Welding J., **7**, (5) 327 to 333 (1960).

240. GHEORGHIU, A.: Melting of Particle Injected into a Plasma Jet. Revu. Roumaine Physi., **14**, (4) 327—336 (1969).

241. GIANNINI, G. M.: The Plasma Jet and Its Application. Paper Presented at the High Intensity Arc Symposium, June, 1957.

242. GIANNINI, G. M.: The Plasma Jet. Sci. Amer., **107**, (2) 80—88 (1957).

243. GIEMZA, C. J., and W. B. HUNTER: Structural Heat Shield for Reentry and Hypersonic Lift Vehicles (High-Temperature Composite Structure). Part I, Volume I, Wright-Patterson Air Force Base, TDR 64-267, U.S. Air Force, 1965.

244. GILLIGAN, J. E., M. E. SIBERT, and T. A. GREENING: Passive Thermal Control Coatings. Lockheed Missiles and Space Co., Palo Alto, California, Defense Documentation Center, AD-602 894 (No Date).

245. GILMAN, W. S., P. W. SEABAUGH, and D. B. SULLENGER: Stable Substoichiometric Cerium Oxide Formed in an Air Plasma. Sci., **106**, (7) 1239 (1968).

246. GLADKOVSKII, V. A., and M. L. ZINSHTEIN: Mekhanicheskie Svoistva Vysokoprochnykh Materialov S Pokrytiiami, Poluchennymi Metodom Plazmennogo Napyleniia (Mechanical Properties of High-Strength Materials with Coatings Obtained by the Plasma Spray-Coating Method). Termoprochnost Materialov I Kon Strukivnykh Elementov, Kiev 1967.

247. GLAZER, F. W.: Fortschrittsber. über Cermets. Planseeber. Pulvermet., (2) 59—70 (1954).

248. GLEASON, F. R.: Miniature Thin-Film Inductors. Motorola, Inc., Defense Documentation Center AD-277 674, 1962.

249. GOETZEL, C. G., and P. LANDLER: Refractory Coatings for Tungsten. New York University, Defense Documentation Center AD-258 574, 1961.

250. GOHEEN, J. L., and R. UNGER: Arc Sprayed Gradated Coatings. Progress Reports, Contract No. AF 33 (616)-7323, U.S. Air Force, 1960—1961.

251. GORDON, G. M.: Tungsten and Rocket Motors. Stanford Research Institute, Defense Documentation Center, AD-265 023 (No Date)

252. GRAHAM, J. W., and ZIMMERMANN: Cermets in Jet Engines. Metal Progr., **73**, (4) 108 to 111 (1958).

253. GRAHAM, J. W., and W. HALL: Protective Coatings for Molybdenum Alloys. Navy Bureau of Weapons, Contract No. as 59-60260 C, U.S. Navy, 1960.

254. GRAVER, C. W., F. D. LOOMIS, and I. MOCKRIN: Crystal Growth of Fluoaluminates. Pennsalt Chemicals Corp., Defense Documentation Center, AD-278 465 (No Date).

255. GREEN, J. J., E. SCHLOMANN, A. PALADINO, and J. S. WAUGH: Fine Grain Dense Ferrites. Raytheon Co., Defense Documentation Center AD-424 668, 1963.

256. GREINER, J. W., B. E. KRAMER, and M. A. LEVINSTEIN: The Effect of Arc Plasma Deposition on the Stability of Nonmetallic Materials. Navy Bureau of Aeronautics, Noas 60-6076-C, U.S. Navy, 1960.

257. GRENIS, A. F., and A. P. LEVITT: Infrared Radiation of Solid Refractory Materials. Amer. Ceram. Soc. Bull., **44**, (11) 901—906 (1965).

258. GRIFFITH, J. S., and S. W. BRADSTREET: Solution Ceramics . . . New Fields in Coatings. Ceram. Ind., **63**, (4) 77 (1954).

259. GRIFFITHS, H.: Industrial Uses of the Plasma Arc. Brit. Welding J., **10**, (11) 546—551 (1963).

260. GRISAFFE, S. J., and W. A. SPITZIG: Metallurgical Bonding of Plasma Sprayed Tungsten on Hot Molybdenum Substrate. NASA-TN-D-2510, National Aeronautics and Space Administration, 1964.

261. GRISAFFE, S. J.: Metallurgical Bonding of Plasma Sprayed Tungsten on Hot Molybdenum Substrates. Welding J. Res. Supplement, **43**, (9) 425—427 (1964).

262. GRISAFFE, S. J., and W. A. SPITZIG: Particle-Substrate Bonding of Plasma-Sprayed Materials. NASA TN D-1705, National Aeronautics and Space Administration, 1963.

263. GRISAFFE, S. J., and W. A. SPITZIG: Observations on Metallurgical Bonding Between Plasma Tungsten and Hot Tungsten Substrates. Amer. Soc. Metals., **56**, (3) 618—628 (1963).

264. GRISAFFE, S. J., and W. A. SPITZIG: Analysis of Bonding Mechanism Between Plasma Sprayed Tungsten and a Stainless Steel Substrate. NASA TN-D-2461, National Aeronautics and Space Administration, 1964.

265. GRISAFFE, S. J.: Simplified Guide to Thermal-Spray Coatings. Machine Design, **39**, (8) 174 (1967).

266. HACKMAN, R. J.: Putting Plasma Jets to Work. Tool Manu. Eng., **46**, (3) 85—89 (1961).

267. HAGAR, R. L.: Electromagnetic Shielding Effectiveness Tests on Selected Materials Applied Over Metal-Plastic Interfaces. Naval Weapons Lab., Rept. NWL-TM-W-5/67, U.S. Navy, 1967.

268. HALL, F. E.: Flame-Sprayed Coatings. Prod. Eng., **36**, (25) 59—64 (1965).

269. HALL, R. W.: Protective Coating Materials. NASA 129003-05-03-22, National Aeronautics and Space Administration (No Date),

270. HALL, W. B.: Systems 400 Coating for the Protection of Columbium. Materials Information Memorandum, General Electric Co., Report No. DM-60-97 (No Date).

271. HALL, W. B., and J. O'GRADY: The Development of an Auxiliary Electrode Thermionic Converter. Wright-Patterson Air Force Base, TDR-63-442, U.S. Air Force, 1963.

272. HALLS, E. E.: Sprayed Zinc Coatings. Electroplat. Met. Spray., (7 & 8) 279, 316 (1954).

273. HARWOOD, J. J.: Protecting Molybdenum at High-Temperatures. Materials and Methods, **44**, (6) 8 (1956).

274. HASS, G. H.: The Coatings that Go On the Satellite. Mag. Magnesium., (8) 4 (1957).

275. HASSION, F. X.: Test Results, On Falex Pins Coated by the Plasma-Arc Process. Springfield Armory, Defense Documentation Center, AD-621 064, 1965.

276. HAY, W. S., and J. L. McDANIEL: Densification Techniques for Refractory Metal Powder Forms. General Dynamics, Defense Documentation Center, AD-421 557, 1963, AD-432 820, 1964.

277. HAYES, R. J., and W. H. ATKINSON: Thermal Emittance of Materials for Spacecraft Radiator Coatings. Amer. Ceram. Soc. Bull., **43**, (9) 616—621 (1964).

278. HAYES, G. A.: An Investigation of the Feasibility of Forming Alloy Coatings with a Plasma Jet. Naval Ordnance Test Station, Defense Documentation Center, AD-253 053, 1961.

279. HEADMAN, M., and T. J. ROSEBERRY: Plasma Spraying: What We Know About It. Welding Design Fabric., **36**, (4) 82—87 (1963).

280. HEADMAN, M. L., F. L. PARKINSON, and T. J. ROSEBERRY: Process Development and Evaluation of Plasma Sprayed Beryllium. Western Gear Corp., 1964.

281. HECHT, N. L.: Plasma Sprayed Ceramic-Metal Composite Coatings. Presented at the 68th Annual Meeting of the American Ceramic Society, May, 1966.

282. HECHT, N. L., and G. A. GRAVES: Ceramic Coatings for Erosion Resistance. FRL-TM 30 Picatinny Arsenal, Dover, New Jersey, U.S. Army, 1962.

283. HEDGER, H. J., and A. R. HALL: Preliminary Observations on the Use of the Induction-Coupled Plasma Torch for the Preparation of Spherical Powder. Powder Met., **4**, (8) 65—67 (1961).

284. HEESTAND, R. L., D. E. KIZER, and C. R. RUDERER: Development of a Protective Coating System for Regeneratively Cooled Thrust Chambers. NAS 3-11186, Battelle Memorial Inst., Columbus, Ohio, 1969.

285. HEIL, O.: Particle Bombardment Bonding and Welding Investigation. N 130356, Heil Scientific Labs. Inc., Belmont, California, 1964.

286. HEITMAN, G. H.: Characterization of Plasma-Arc-Sprayed Tungsten. Polaris Propulsion Development. Aerojet-General Corp., Defense Documentation Center, AD-457 420, 1964.

287. HELFRICH, W. E.: Protective Coating for Extended Life of Aircraft Jet Engines Parts. Paper 660310, Society of Automative Engineers, 1966.

288. HELGESSON, C. I.: Phase Determination in Flame-Sprayed Nickel Aluminide Coating. Swedish Institute for Silicate Research, Nature, **209**, (2) 706—707 (1966).

289. HELLUND, E. J.: The Plasma State. New York: Reinhold, 1961.

290. HERZIG, A. J., and J. R. BLANCHARD: Protecting Molybdenum from Oxidation. Metal Progr., **66**, (10) 109 (1954).

291. HESSENBRUCH, W.: Metalle und Legierungen für hohe Temperaturen (Metals and Alloys for High Temperatures). Berlin: Springer, 1940.

292. HESSLER, G.: Das Lichtbogenspritzverfahren und seine praktische Nutzanwendung (Arc Spraying Technology and its Practical Applications). Maschinenmarkt, **66**, (93) 229—232 (1960), **67**, (24) 20—24 (1961).

293. HICKS, W. T., and H. VALDSAAR: Thermoelectric Properties of Selenides and Tellurides of Groups VB and VIB Metals and Their Solid Solutions. Du Pont De Nemours, Defense Documentation Center AD-419 514, 1963.

294. HILDEBRAND, J. F., E. W. TURNS, and F. C. NORDQUIST: Stress Corrosion Cracking in High Strength Ferrous Alloys. General Dynamics Defense Documentation Center AD-423 387, 1963.

295. HILL, R. B.: Use of Flame Sprayed Coatings for Reduction in Initial and Subsequent Repair Costs of Gas Turbine and Components. American Society of Mechanical Engineers, Gas Turbine Conference, New York, March, 1968.

296. HILL, R. J., and E. G. WOLFF: Evaluation of Fabrication Techniques for Metal Matrix Composites. J. Composite Mater., **2**, (7) 405—407 (1968).

297. HILL, V. L.: Tungsten Fabrication by Arc Spraying. J. Metals, **13**, (7) (1961).

298. HIRAKIS, E. C.: Coatings for the Protection of Columbium at Elevated Temperatures. Wright-Patterson Air Force Base, WADC TR 58-545, U.S. Air Force (No Date).

299. HJELM, L. N.: Development of Improved Ceramic Coatings to Increase the Life of XLR-99 Thrust Chamber. Wright-Patterson Air Force Base, TM-62-2, ASRCE, U.S. Air Force, 1962.

300. HOEHNE, K.: Possibilities of the Ultrasonic Testing of Layers of Metal Spray with Respect to Adhesion and Homogeneity. Wright-Patterson Air Force Base, FTD-TT-64-1334, U.S. Air Force, 1965.

301. HOENING, C. L., et al.: High-Temperature Resistant Ceramic Coatings, Ceramic and Metal Ceramic Bodies. Contract No. AF 33 (616)-2307, U.S. Air Force, 1954.

302. HOLDER, S. G., and A. C. WILLHELM: Protective Coatings for Sheet Metals in Supersonic Transport Aircraft. NASA Contract NASR-117 Report 5798-1417-III, National Aeronautics and Space Administration, 1962—1963.

303. HOLGATE, S. M.: Spraywelding. Electroplat. Met. Spray., (6) 239—245, 1954.

304. HOLGATE, S. M.: A Review of the Use of Molybdenum in Metal Spraying. Electroplat. Met. Finishing, **8**, (7) 258—262 (1955).

305. HOLTZ, F. C.: High-Temperature Oxidation Protective Coatings for Vanadium-Base Alloys. Navy Office of Weapons, NOW 61 806, U.S. Navy, 1962.

306. HOOTON, N. A.: Materials Property Data. Bendix Products Corp., Defense Documentation Center AD-266 937, 1961.

307. HOPKINS, V., M. LAVIK, W. CLOW, C. BOLZE, and R. HUBBELL: Development of New and Improved High-Temperature Solid Film Lubricants. Wright-Patterson Air Force Base, TDR-64-37, U.S. Air Force, 1966.

308. HOPKINS, V., R. D. KREMITH, and J. W. ROSENBERRY: Solid Lubricant Coatings Applied by Plasma Spray. Amer. Ceram. Soc. Bull., **47**, (9) 813—818 (1968).

309. HOPKINS, V., R. HUBBELL, and R. KREMITH: Plasma Spraying—A New Method of Applying Solid Film Lubricants. Amer. Soc. of Lubri. Eng., Toronto, Canada, May, 1966.

310. HOUGHTON, E. K.: Apparatus for Coating with Glass or Enamel. U.S. Patent 1, 586, 990 (1926).

311. HRACH, G. C.: Preparation of Space Hardware by Plasma Flame Throwing Techniques. ASTM Paper 26, Amer. Soc. of Testing Mater., October, 1962.
312. HUBBEL, W. G.: Ceramic Coated Exhaust Systems. Aircraft Prod., **13**, (15) 351 (1951).
313. HUFFADINE, J. B., and A. G. THOMAS: Flame Spraying as a Method of Fabricating Dense Bodies of Alumina. Powder Met., **7**, (14) 290—299 (1964).
314. HUMBARGER, F. F.: Effective Tooling for Flame Plating. Machinery, **70**, (6) 132—136 (1963).
315. HUMINIK, J.: High-Temperature Inorganic Coatings. New York: Reinhold Publishing Corp., 1963.
316. HUTZLER, J. R.: Research and Development Work on Development of Thin Organic Rolled Film Capacitor. Linde Div. Union Carbide Corp., Defense Documentation Center AD-623 905, 1965.
317. INGHAM, H. S., and A. P. SHEPARD: Metallizing Handbook. Volume I, II, III, Long Island, N.Y.: Metco Inc., 1951.
318. INGHAM, H. S.: Flame Sprayed Protective Coatings, Made from Refractory Materials. Mater. Prot., **1**, (1) 74—78 (1962).
319. INGHAM, H. S.: Sprayed Metals as a Base for Paints. Corro., **13**, (4) 252—256 (1957).
320. INGHAM, H. S.: Developments in Sprayed Metal Coatings. Metal Working Prod., **100**, (6) 1521 (1956).
321. INGHAM, H. S.: Flame-Sprayed Coatings. Chap. 15 of Composite Engineering Laminates, Ed. DIETZ, A.G.H., Cambridge, Mass.: M.I.T. Press, 1970.
322. INGHAM, H. S., and A. P. SHEPARD: Evaluation Methods and Equipment for Flame-Sprayed Coatings. Research Laboratory Report # 106. Long Island, N.Y.: Metco Inc., 1963.
323. INGRAHAM, J. M., and M. M. MARDIROSIAN: Feasibility Study of a Multilayer Flame-Spray Coated and a Brazed Platelet-Coated Armor Plate as a Defense Against Heat Ammunition. Technical Report # WAL TR 161.86/2, Watertown Arsenal Labs., U.S. Army, 1960.
324. JACKMAN, P.: Before You Plasma Spray. Amer. Machinist, **108**, (3) 56—57 (1964).
325. JACKSON, C. M., and A. M. HALL: Surface Treatments for Nickel and Nickel-Base Alloys. Battelle Memorial Inst., Defense Documentation Center AD-634 076, 1966.
326. JANOWIECKI, R. J., G. F. SCHMIDT, JR., and M. C. WILLSON: Plasma Sprayed High-Temperature Polymeric Coatings. SAMPE J., **4**, (41) 40—49 (1968).
327. JANOWIECKI, R. J., M. C. WILLSON, and D. H. HARRIS: Plasma Sprayed Thermoelectric Materials. Monsanto Research Corp., Defense Documentation Center, AD-832 678, 1968.
328. JANOWIECKI, R. J., et al.: High-Temperature Thermoelectric Generator. Wright-Patterson Air Force Base, TDR 62-896, Part 2, U.S. Air Force, 1963.
329. JOHNSON, R. L.: Flame Sprayed Metallized Aluminum for Protection of Carbon Steel Parts in the Titan B Missile Silo. Martin Marietta Corp., Defense Documentation Center, AD-292 385, 1961.
330. JOHNSON, R. L., and W. M. WHEILDON: Faster Plasma Coating. Mater. Design Eng., **56**, (6) 16—17 (1962).
331. JONES, R. A., and L. T. FUSZARA: Problems of Utilizing Ceramics in Aircraft Power Plant Construction. Amer. Ceram. Soc. Bull., **32**, (1) 7—9 (1953).
332. KALSING, H.: Keramische Stoffe für Raketenantriebe. Sprechsaal Keram., Glas, Email, **89**, (12) 414—416 (1956).
333. KATTS, N. V., E. V. ANTOSHIN, D. G. VADIVASOV, G. D. VOLPERT, and L. M. KAMIONSKII: Spray Metallizing. Army Foreign Science and Technology Center, FSTC-HT-23-10-68, U.S. Army, 1966.

334. Katz, N. N.: Einige Probleme der Lichtbogenspritztechnik in der Sowjetunion (Several Problems of Arc Spraying Technology in the Soviet Union). Schweißtech., **7**, (4) 137 to 141 (1957).

335. Katz, N. N., and E. M. Linnik: Elektrometallspritzen (Electric Metal Spraying). Moskow: Zelchosgis Verlag, 1953.

336. Kenderi, T.: Gefüge und Schutzwirkung gespritzter Metallschichten (Structure and Protective Characteristics of Sprayed Metal Layers). Metalloberfläche, **13**, (5) 133—134 (1959).

337. Kennedy, A. J.: A Study Program on Cesium Vapor Filled Thermionic Converters having High Vacuum Work Function Emitter Materials. Martin Marietta Corp., Defense Documentation Center AD-608 554, 1964.

338. Kennedy, A. J., and D. S. Trimmer: The Performance of Ruthenium as an Electrode in a Thermionic Converter. Martin Pattern Co., Defense Documentation Center, AD-465 704, 1964.

339. Kinas, E. N.: Titanium Alloy for T-109 Medium Tank Track Development of Processing Procedures and Manufacturing Techniques. Water Pollution Research Board Watford England, Defense Documentation Center AD-264 834, 1961.

340. King, B. W.: Ceramic Coated Metals for Industry. Battelle Tech. Rev., **3**, 39—42 (1954).

341. King, R. M.: Ceramic Coatings for Metals. Defense Documentation Center AD-19325, 1953—1954.

342. Kirner, K.: Plasmagespritzte, elektrisch isolierende Keramikschichten (Plasma Sprayed Electrically Insulating Ceramics Coatings). Physikalische Laboratorien, Stuttgart West Germany, **2**, (5) (1968).

343. Kizer, D. E., and D. E., Lozier: Powder Metallurgy Review of Recent Developments. Defense Metals Information Center, N 64-33509, 1964.

344. Kizer, D. E., and D. E. Lozier: Powder Metallurgy Review of Recent Developments. Defense Metals Information Center, N 65-24355, 1965.

345. Klein, L.: Evaluation of Ablative and Insulation Coatings for Missile Structure. Martin Marietta Corp., Defense Documentation Center AD-453 120, 1960.

346. Klopp, W. D., and C. F. Powell: Development of Protective Coatings for Tantalum-Base Alloys. Wright-Patterson Air Force Base, TR 61-676, U.S. Air Force (No Date).

347. Knanishu, J.: Galvanic Protection by Metal Spray Method. Rock Island Arsenal, Illinois, Defense Documentation Center, AD-412 499, 1963.

348. Koch, J. C.: Plasma-Sprayed Uranium Carbide Coating of Graphite Crucibles. Amer. Ceram. Soc. Bull., **49**, (5) 549—550 (1970).

349. Koch, H., and I. Adams: Einfluß der Arbeitsbedingungen auf die Eigenschaften gespritzter Stahlschichten (Effect of Working Conditions on the Properties of Sprayed Steel Layers). Schw. U.-Schn., **5**, (4) 131—142 (1953).

350. Koenig, R. F.: Manufacturing Techniques for Application of Erosion Resistant Coatings to Turbine Engine Compressor Components. Defense Documentation Center, AD-837 238, 1968.

351. Kordes, E.: Kristallchemische Untersuchungen über Silicium Oxydverbindungen mit spinellartigem Gitterbau über γ-Fe$_2$O$_3$. Z. Krist., **91**, 193—228 (1935).

352. Koubek, F. J., H. A. Perry, and I. Silver: Polaris Materials Program I Literature Survey II Thermal and Mechanical Properties of Ceramics, Cermets and Metals. Nav Ord Report 6056, U.S. Naval Ordnance Laboratory, U.S. Navy, 1958.

353. Koubek, F. J., and A. R. Timmins: High-Temperature Testing of Ceramics for Reentry Body Applications. Nav Ord Report 6298, U.S. Naval Ordnance Laboratory, U.S. Navy, 1959.

354. Kozlovskiy, A. L.: Protective Flame-Spray Coatings. Defense Documentation Center, AD-405 249 (No Date).

355. KRAMER, B. E.: Development of Arc Spraying Processes and Materials for Solid Rocket Nozzles. NORD 18119, U.S. Navy, 1961.
356. KRAMER, B. E., M. A. LEVINSTEIN, and J. W. GRENIER: The Effect of Hot Plasma Deposition on the Stability of Non-Metallic Materials. General Electric Co., Defense Documentation Center, AD-264 602, 1961.
357. KRASKA, I. R., and E. KUBIAK: Investigation of Nondestructive Test Methods for Metallized Tank Engine Cylinders. General American Transportation Corp., Niles, Ill., Technical Report, December, 1966, February, 1968.
358. KRASNOV, A. N., and S. I., SHARIVKER: Plazmennye Pokrytiaa (Plasma Coating). Transactions Vysokotemperaturnye Pokrytiia, Seminar Po Zharostoikim Pokrytiiam, Leningrad, U.S.S.R., 1965.
359. KREIDER, K. G.: Services and Materials Necessary to Develop a Process to Produce Fibrous Reinforced Metal Composite Materials. United Aircraft Corp., Defense Documentation Center AD-478 306, 1965—1966.
360. KREIDER, K. G., and G. R. LEVERANT: Boron Fiber Metal Matrix Composites by Plasma Spraying Spraying. Wright-Patterson Air Force Base, TR-66-219, U.S. Air Force, 1966.
361. KREIDER, K. G., and G. R. LEVERANT: Born Aluminum Composite Fabricated by Plasma Spraying. Advanced Fibrous Reinforced Composites, 10, F1—F9, Science of Advanced Materials and Process Engineering Series, N. Hollywood, California: Western Periodicals Co., 1966.
362. KREIDER, K. G., R. D. SCHILE, E. M. BREINAN, and M. A. MARCIANO: Plasma Sprayed Metal Matrix Fiber Reinforced Composites. Wright-Patterson Air Force Base, TR-68-119, U.S. Air Force, 1968.
363. KREKELER, K., and K. STEINEMER: Metallspritzen. Werkstattbücher, No. 93, Berlin-Göttingen-Heidelberg: Springer-Verlag, 1952.
364. KREMITH, R. D., and J. W. ROSENBERRY: Plasma Spray Application of Plastic Materials Including Thermosetting Epoxies and Polyesters and Thermoplastic Polyethylene and Polyamides. Advances in Structural Composites, 12, SAMPE Science of Advanced Materials and Process Engineering, N. Hollywood, California: Western Periodicals Co., 1967.
365. KRETZSCHMAR, E.: Das Metallspritzverfahren (Metal Spraying Technology). Halle/Saale: VEB Karl Marhold Publishers, 1953.
366. KRIER, C. A.: Coating for the Protection of Refractory Metals. Defense Metals Information Center, Battelle Memorial Institute Defense Documentation Center, AD-271 384, 1961.
367. KULAGIN, I. D., and A. V. NIKOLAEV: The Arc Plasma Jet as a Heat Source in the Working of Materials. Welding Prod., 5, (9) 1—11 (1959).
368. LAKE, F. N., E. J. BREZNYAK, and G. S. DOBLE: Tungsten Forging Development Program. Thompson Ramo Wooldridge Inc., Defense Documentation Center, AD-258 195, 1961.
369. LASZLO, T. S.: Mechanical Adherence of Flame-Sprayed Coating. Amer. Ceram. Soc. Bull., 40, (12) 751—755 (1961).
370. LAWRIE, W. E., et al.: Non-destructive Methods for the Evaluation of Ceramic Coatings. Wright-Patterson Air Force Base, WADD TR-61-91, Part II, Part III, Part IV, U.S. Air Force (No Date).
371. LEEDS, D. H.: Some Observation of the Interface Between Plasma Sprayed Tungsten and 1020 Steel. Aerospace Corp., Defense Documentation Center AD-803 286, 1962.
372. LEEDS, D. H.: The Interface Between Plasma-Sprayed Tungsten and 1020 Steel and the Mechanism of the Plasma-Sprayed Coating Bond. Metallurgy and Ceramic Materials Sciences Laboratory, Aerospace Corp. (No Date).

373. LEEDS, D. H.: Materials and Structure Program, High Mach Phase III Arc Plasma Spray Investigations. Aerospace Corp., TDR-930 2240, U.S. Air Force, 1962.
374. LEEDS, D. H.: Coatings on Refractory Metals. Chapt. 7, Ceramics for Advanced Technologies, J. E. HOWE, W. C. RILEY, Ed., New York: John Wiley and Son, 1965.
375. LEGGETT, H., and R. L. JOHNSON: Development and Evaluation of Insulating Type Ceramic Coating. Wright-Patterson Air Force Base, TR-59-102 P 2, U.S. Air Force, 1960.
376. LENTZ, J. K.: Research and Development of Unitized, Miniaturized Relay. Philips Labs., Inc., Defense Documentation Center, AD-250 963 (No Date).
377. LESZYNSKI, S. W.: The Development of Flame Sprayed Sensors. Boeing Co., Defense Documentation Center, AD-283 958, 1961.
378. LEVINSTEIN, M. A., A. EISENLOHR, and B. E. KRAMER: Properties of Plasma Sprayed Materials. Welding J., **40**, (1) 8 s—13 s (1961).
379. LEVINSTEIN, M. A.: Properties of the Refractory Metals Sprayed Under Controlled Environment, J. Metals, **14**, (2) 137 (1962).
380. LEVINSTEIN, M. A., et al.: Properties of Plasma Sprayed Materials. Wright-Patterson Air Force Base, TDR-62-201 and TR-60-654, U.S. Air Force, 1962.
381. LEVINSTEIN, M. A., and C. E. JOHNSON: Properties of Plasma Sprayed Materials. Quarterly Reports, AF 33 (616)-6376, Task 73810, January-March, 1961, April-June, 1961, July-September, U.S. Air Force, 1961.
382. LEVINSTEIN, M. A.: Plasma Spraying-State of the Art. Report AF 33 (616)-6376, Task 7381, U.S. Air Force, 1961.
383. LEVINSTEIN, M. A., ed.: Recent Advances in Arc-Plasma Metallizing. Metal Finishing J., **6**, (72) 467—474 (1960).
384. LEVITT, A. P., and M. LEVY: Flame-Sprayed Metallic and Ceramic Coatings for Army Applications. Presented at 3D International Metallization Conference, Madrid, May, 1962, Army Materials Research Agency, MS-64-01, U.S. Army, 1962.
385. LEVY, A. V.: Ceramic Coatings for Insulation. Metal Prog., **75**, (3) 86—89 (1959).
386. LEVY, M.: Evaluation of Flame-Sprayed Coatings for Army Weapons Applications. Amer. Ceram. Soc. Bull., **42**, (9) 498—500 (1963).
387. LEVY, M.: Trip Report to the III International Metallization Conference, Madrid, Spain. Watertown Arsenal Laboratories, U.S. Army, 1962.
388. LEVY, M., and A. P. LEVITT: Application of Flame-Sprayed Coatings at Watertown Arsenal. WAL TR-371.1/1, U.S. Army, 1961.
389. LEVY, M., and D. J. SELLERS: A Study of the Effect of Heat Treatment on the Microstructure and Properties of Sprayed Molybdenum. Army Materials Research Agency, TR-64-50, U.S. Army, 1964.
390. LEVY, M., G. N. SKLOVER, and D. J. SELLERS: Adhesion and Thermal Properties of Refractory Coating-Metal Substrate Systems. Army Materials Research Agency, TR-66-01, U.S. Army, 1966.
391. LEWIS, W. J.: Coatings for Advanced Thrust Chambers. NASA-CR-72604, National Aeronautics and Space Administration, 1968.
392. LIEBERT, C. II.: Spectral Emittance of Aluminium Oxide and Zinc Oxide on Opaque Substrates. NASA TN D-3115, National Aeronautics and Space Administration, 1965.
393. LOHRIE, B.: Plasma Jets Headed for Production Roles. Steel, **147**, 110—112 (1960).
394. LOMATZ, J. P., and D. H. LEEDS: Douglas Process Standard 9,500 Flame Sprayed Ceramic Coatings. Douglas Aircraft Co., Inc., 1957.
395. LONG, J. D., and J. PSAROUTHAKIS: Investigations Using Segmented Collector Thermionic Converters. Office of Naval Research Report 099-363, U.S. Navy, 1966.
396. LONGO, F. N.: Techniques for Improving Plasma-Sprayed Tungsten. Society of Aerospace Material and Process Engineers, National Symposium on Materials for Space Vehicle Use, Seattle, Washington, Volume I, 1963.

397. Longo, F. N.: Metallography of Flame Sprayed Coatings of Metco 404, Nickel Aluminide Powder. Research Lab. Report 107, Long Island, N.Y.: Metco Inc., 1963.

398. Longo, F. N.: Metallurgy of Flame Sprayed Nickel Aluminide Coatings. Welding Res. Supplement, **47**, (2) 665—695 (1966).

399. Ludke, G., and J. Sternkopf: Flame-Spraying of Ceramic Coatings. Wright-Patterson Air Force Base FSTC-381-T 65-271, U.S. Air Force, 1964.

400. Lutz, O.: Werkstoffe für feuergerührte Bauteile im Strahltriebwerken. Luftfahrttechn., **1**, 118—123 (1955).

401. Mac-Dowall, K.: Preliminary Evaluation of Flame Sprayed Aluminum on H-11 Steel. TFD-59-1162, North American Aviation, Inc., Calif., 1959.

402. Machenschalk, R.: Aufgespritzte Molybdänüberzüge (Sprayed Molybdenum Coatings). Planseeber. Pulvermet., **4**, (4) 80—84 (1956).

403. Mackay, T. L., and A. N. Muller: Plasma Sprayed Dielectric Coatings for Heat Sinks in Electronic Packaging. Amer. Ceram. Soc. Bull., **46**, (9) 833 (1967).

404. Malik, M. P.: Die Anwendung des Plasma Metallspritzverfahrens bei der Reparatur von Flugzeugbauteilen aus Al- und Mg-Legierungen (Use of Plasma Metal Spray Coating in Repairing Aircraft Components Made from Al and Mg Alloys). Luftfahrttech. Raumfahrttech., **15**, 188—191 (1969).

405. Malinofsky, W. W., and R. W. Babbitt: Fine-Grained Ferrites, III, $Ni_{(1-x)}Co_{(x)}Fe_2O_4$, RF Properties. J. of Appl. Phys., **35**, (3) Part 2, 1012—1014 (1964).

406. Manning, H. E.: Spray Metallizing of Plastic Laminates-Process Specification for. Defense Documentation Center, AD-817 435 L, 1967.

407. Mansford, R. E.: Sprayed Aluminum and Zinc in Corrosive Environments. Corrosion Tech., **3**, (10) 314 (1956).

408. Manuel, L.: Properties and Applications of the Flame-Sprayed Refractory Coatings. Metal Finishing J., **4**, (44) 313—316 (1958).

409. Marnock, K.: Research on High-Temperature Oxidation Resistant Hafnium-Tantalum Systems. Contract AF 33 (615)-1628, Report 25142, U.S. Air Force, 1964.

410. Marshal, P. H.: Metal Spraying. J. Inc. Plant. Eng., (8) 299, 309 (1953).

411. Maruo, H., and M. Okada: New Plasma Spraying and Its Application. Brit. Welding J., **15**, 371—386 (1968).

412. Marynowski, C. W., F. A. Halden, and E. P. Farley: Variables in Plasma Spraying. Electrochem. Tech., **3**, (3—4) 105—115 (1965).

413. Mash, D. R.: Plasma-Arc Spraying of Space-Age Materials. Western Machinery Steel World, **3**, (4) 48—53 (1962).

414. Mash, D. R.: Plasma-Arc Spraying of Refractory Metals. Canadian Machinery Metalwork., **73**, (8) 81—83 (1962).

415. Mash, D. R., N. E. Weare, and D. L. Walker: Process Variables in Plasma-Jet Spraying. J. Metals, **13**, (7) 473—478 (1961).

416. Mason, C. R., J. D. Walton, and C. A. Murphy: Fused Silica Rocket Nozzles. Georgia Inst. of Tech., NOrd-18564, U.S. Navy, 1960.

417. Mason, C. R., J. D. Walton, C. A. Murphy, and A. T. Sales: Investigation of High-Temperature Resistant Materials. Georgia Inst. of Tech., NOrd-15701, U.S. Navy, 1961.

418. Mathauser, E. E.: Ceramic-Metal Composites for Structural Applications. NASA-129-03-09-01-23, National Aeronautic and Space Administration (No Date).

419. Matting, A.: Metal Spraying from Gas Flame to Plasma Jet. Brit. Welding J., **4313**, (9) 526—532 (1966).

420. Matting, A.: Schweißtechnische Oberflächenbehandlung zur Herstellung von Überzügen (Surface Preparation Based on Surface Preparation Techniques for Welding for Producing Coatings). Maschinenbau, Der Betrieb, **14**, (23/24) 683—686 (1935).

421. Matting, A., and K. Becker: An Experimental Investigation of the Metal Spraying Process. Electroplat. Metal Finish., **8**, 101—103, 143—145 (1955), and **9**, 85, 88, 126—128, 147—148 (1956). [Also in Schw. U. Schn., **6**, (4) 127—141 (1954).]

422. Matting, A., and K. Becker: Investigations Into the Metal Spraying Process. Eng. Digest., (8) 309 (1954).

423. Matting, A., and W. Raabe: Der Aufbau von Metallspritzschichten (The Formation of Sprayed Metal Layers). Schw. U. Schn., **8**, (10) 369—374 (1956).

424. Matting, A., and H. D. Steffens: Die Wissenschaftlichen Grundlagen des Metall-spritzens (The Scientific Principles of Metal Spraying). Mitt. Forschungsges., Blechver-arb., (16/17), 220—230 (1962).

425. Matting, H. A., and H. D. Steffens: Haftung und Schichtaufbau beim Lichtbogen und Flammspritzen (Adhesion and Layer Formation for Arc and Flame Spraying). Metall., **17**, (6) 583—593, (9) 905—922, and (12) 1213—1227 (1963).

426. Matting, A., and H. D. Steffens: Beitrag zur Erforschung des Lichtbogen-Metall-spritzprozesses (Contribution to the Research on Arc Metal Spraying Process). Zeit-schrift für Metallkunde, **53**, (2) 138—144 (1962).

427. McCullum, D. E., and N. L. Hecht: Plasma Sprayed Coatings for Thermal Protection of Rocket Sled Components. Engineering Test Memorandum 12, University of Dayton, Contract AF 33 (615)-1312, U.S. Air Force, 1966.

428. McDaniels, D. L., R. W. Jeck, and J. W. Weeten: Metals Reinforced with Fibers. Metal. Progr., **78**, (6) 117 (1960).

429. McDermott, W., and R. Dickinson: The Influence of Fuel Gases on the Spraying of Metal Powders. Electroplat. Met. Spray., (8) 309 (1953).

430. McGeary, T. C.: Engineering Applications for Flame Plating. Metal. Progr., **87**, (1) 80—86 (1965).

431. McLean, W. J.: Relative Benefit Afforded by Coating a Heat Sink with a Thin Ceramic Layer. NavOrd 6694, U.S. Navy, 1959.

432. Mehegan, P.: Rokide Z Thrust Chamber Nozzle Coatings. Rocketdyne Report CER 0120-7301, 1959.

433. Merry, J. D., and C. H., Vandracek: Three Uses of Flame Sprayed Al_2O_3. Materials Laboratories Westinghouse Electric Corp., 1963.

434. Meuer, M.: Improvements in and Relating to Coating Heat Resisting Articles by Spray-ing with Enamels, Glazes and Like Substances. Patent Specification 179, 216 (1922).

435. Meyer, H.: On the Flame-Spraying of Aluminum Oxide. Base Mater. Corrosion, **11**, (10) 601—616 (1960).

436. Meyer, H.: Flame-Spraying of Alumina. Royal Aircraft Establishment Defense Doc-umentation Center, AD-258 168 (No Date).

437. Meyer, H.: Das Verhalten von Pulvern im Plasmastrahl. Ber. Dtsch. Keram. Ges., **39**, (H2) 115—124 (1963).

438. Meyer, H., and A. Dietzel: Das Flammenspritzen von keramischen Überzügen (The Flame Spraying of Ceramic Coatings). Ber. Dtsch. Keram. Ges., **37**, (4) 136—141 (1960).

439. Meyer, R. J.: Evaluation of Rokide Z Coatings on Nozzle. Report 038-086 McDonnell, St. Louis, Missiouri (No Date).

440. Michael, H. J.: Aircraft Coating, Testing, and Evaluation. North American Aviation, Inc., Defense Documentation Center, AD-814 759, 1967.

441. Miller, H. S.: Practical Hard Facing with Fused Self Fluxing Metallized Coatings. Welding J., **34**, (3) 214 (1955).

442. Miller, R. C., and A. W. Brunet: Surface Preparation and Fatigue in Metal Spraying. Electroplat. Metal Spray. (10) 393 (1954).

443. Mock, J. A.: Plasma Arc-Torch Fabricates Tough Materials. Mater. Design Eng., **49**, (3) 133—134 (1959).

444. MOCK, J. A.: Flame Sprayed Coatings. Mater. Design Eng., **63**, (2) 89—104 (1966).

445. MONROE, R. E.: New Joining Processes for Uncommon Materials. Mech. Eng., **83**, (8) 79—80 (1961).

446. MONTGOMERY, E. T., J. E. ANTHONY, and K. P. PAOLETTI: Cermet Coatings for Gas Turbines Burning Residual Fuel Oils. Ohio State University, Universal Research Foundation, Defense Documentation Center, AD 47 724, 1954.

447. MOORE, D. G., A. G. EUBANKS, H. R. THORNTON, W. D. HAYES, and A. W. GRIGLER: Studies of the Particle Impact Process for Applying Ceramic Cermet Coatings. Defense Documentation Center, AD-266 381, 1961.

448. MOORE, D. G., W. D. HAYES, and A. W. GRIGLER: Velocity Measurements of Flame-Sprayed Aluminum Oxide Particles. Phase I, Contract AF 33 (616) 59-19, Project #8 (88—702), U.S. Air Force, 1959.

449. MOORE, V. S., and A. R. STETSON: Evaluation of Coated Refractory Metal Foils. Wright-Patterson Air Force Base, TDR-63-4006, U.S. Air Force, 1963.

450. MORENOV, I. A., and A. V. PETROV: Opredelnie Skorosti Chastits Napyliaemogo Materiala Metodom Skorostnoi Kinos Emki (Determination of the Particle Velocity of Sprayed Material by the Method of High-Speed Cinematography). Poroshkvaia Met., **7**, (9) (1967).

451. MORETON, R., and R. W. GARDINER: The Protection of Molybdenum from Oxidation at High-Temperatures with Zirconia/Glass Coatings. TN CPM 32, Royal Aircraft Establishment Farnborough, England, 1963.

452. MOROZOV, I. A.: Ostatochnye Napriazheniia Pri Plazmennom Napylenii. Mashinostroenie, (5) 111—115 (1969).

453. MORRIS, R. J.: Development of High Strength Materials for Solid Rocket Motors. General Electric Co., NOrd-18119, U.S. Navy, 1960.

454. MOSBY, H. V.: Plasma Processes. Encyclopedia of Engineering Materials and Processes (pp. 479—480). Ed. by H. R. CLAUSER, New York: Reinhold, 1963.

455. MOSBY, H. V.: Plasma-Jet Coating: Ultra-High-Temperature Applications for Rocket Motors. Aircraft Prod., **23**, (6) 206—208 (1961).

456. MOSS, A. R., and W. J. YOUNG: The Role of Arc-Plasma in Metallurgy. Powder Met., **7**, (14) 261—288 (1964).

457. MOSS, M., and D. M. SCHUSTER: Mechanical Properties of Dispersion Strengthened Spray-Quenched Al-V Alloys. ASM Trans. Quart., **62**, 201—205 (1969).

458. MOSS, M.: Dispersion Hardening in Al-V by Plasma-Jet Spray Quenching. Acta Met., **16**, 321—326 (1968).

459. MULLER, W. C., and A. K. WOLFF: Study of Methods to Control Grain Size, Purity and Texture in Beryllium and Beryllium Alloys. Nuclear Metals Division, Textron Inc., NMI-8001. 6 NOW-65-0084, U.S. Navy, 1966.

460. MURPHY, C. A., N. E. POULOS, and J. D. WALTON: Spray-On Refractory Coatings System Considerations. Paper presented at National Pyro-Metallurgical Program, AIME, 1963.

461. NAGLER, R. G.: Application of Spectroscopic Temperature Measuring Methods to Definition of a Plasma Arc Flame. Technical Report # 32—66, Jet Propulsion Lab. Cal. Tech., 1961.

462. NAJMON, R. A.: Resistors, Fixed, Precision High-Temperature Radiation Resistant. Mallory and Co., Defense Documentation Center, AD 262 63, 1960.

463. NASH, D. R., N. E. WEARE, and D. L. WALKER: Process Variables in Plasma-Jet Spraying. J. Metals, **13**, (7) 473 (1961).

464. NELSON, C. E.: Method of Applying Protective Coatings. U.S. Patent 1, 566, 911, 1952.

465. NELSON, C. E.: High-Temperature Resistant Ceramic Coatings. Wright-Patterson Air Force Base, WADC 56-139, U.S. Air Force, 1955.

466. NESSLER, C. G., and J. R. PALERMO: Plasma Arc Coatings. Mater. Design Eng., **55**, (6) 109—113 (1962).

467. NEWCOMER, R.: Precoats for Adhesion of Sprayed Aluminum Coatings. McDonnell Aircraft Corp. A 241, Contract AF 33 (657)-11215, U.S. Air Force, 1953.

468. NEWKIRK, H. W.: Continuous Arc Fusion of Uranium Dioxide Powder. General Electric Co., Nuc. Sci. Abstr., (4) 1199 (1959).

469. NIEDERGESON, B. F.: Method of Producing Clear Vitreous Silica. General Electric Co., U.S. Patent 1, 896, 163, 1932.

470. NIEHAUS, W. R., J. E. SHROUT, and R. G. ANDERSON: Structural Heat Shield for Re-Entry and Hypersonic Lift Vehicles (High-Temperature Composite Structure) Test Evaluation. Wright-Patterson Air Force Base, TDR-64-267-Part 2, Volume 1, U.S. Air Force, 1966.

471. NIKOLAYEVA, T. N., V. G. KURYATNIKOVA, and N. S. KUDRYAVTSEVA: Anticorrosive Fluorplastic — 3 and Fluoroplastic-3M Coatings. Foreign Technology Div., U.S. Air Force, Defense Documentation Center, AD-405 242, 1962.

472. NIKOLAYYEV, G. A., and A. I. AKULOC: Welding. Foreign Technology Division U.S. Air Force, Defense Documentation Center, AD-260 686 (No Date).

473. NIMVITSKAYA, T. A.: Deposition of Refraction Coatings with the Use of Plasma. Wright-Patterson Air Force Base, FTD-HT-23-176-69, U.S. Air Force, 1969.

474. NOLTING, H. J.: Oxidation Resistant High-Temperature Protective Coatings for Tungsten. Thompson Ramo Wooldridge Inc., TM-3890 Quarterly Progress Report No. 3, November, 1965.

475. NORTHROP, H.: Fabrication of Pyrolytic Graphite Solid Rocket Nozzle Components. General Electric Co., NOrd-19119, U.S. Navy (No Date).

476. O'BRIEN, R. L.: Plasma Arc; New Fabricating Tool. S.A.E. J., **69**, (7) 84—86 (1961).

477. OECHALE, S. T.: Metal Spraying-Development and Application. Metal Finishing J., **55**, (66) 71—76 (1957).

478. OKADA, M., et al.: Fundamental Researches on Plasma Jet and Its Application. Tech. Rep. Osaka Univ., **10**, (384) 209—219 (1960).

479. OLD, A. R.: Metal Spraying. J. Oil Color Chemists Assoc., **35**, (379) 20 (No Date).

480. OLIEVSKII, M. I.: Povedenie V Plazme Karbidov Tsirkoniia I Niobiaa (Behavior of Zirconium and Niobium Carbides in a Plasma). Poroshkovaia Metall., **8**, 74—81 (1968).

481. OLIEVSKII, M. I.: Plotnost' Keramicheskikh Plazmennykh Polrytii (Density of Ceramic Plasma Coatings). Poroshkovaia Metall., **9**, 38—43 (1969).

482. ORBACH, H. K.: Ceramic Uses of Plasma Jet. Ceram. Ind., **79**, (5) 72—75 (1962).

483. OXX, G. D.: Which Coatings at High-Temperature. Prod. Eng., **29**, 61 (1958).

484. PALENA, M.: Flame Sprayed Ceramic Coating Techniques. Thiokol, Reaction Motors Division Report RMD-9368 F, 1964.

485. PALERMO, J. R., and C. C. POLTER: Plasma Spraying Present and Future. Internal Report, Thermal Dynamics Corp., Lebanon, New Hampshire (No Date).

486. PANKRATOV, B. M.: Nekotorye Metody Zashchity Konstruktsionnykh Materialov (Some Methods of Protecting Structural Materials). Transactions Vysokotemperaturnye Pokrytiia, Seminar Po Zharostoikim Pokrytiiam Leningrad, U.S.S.R., 1965.

487. PASTERICK, N. R., and G. W. FISHER: Plasma Metallizing a Compressor Case. Amer. Machinist/Metalworking, **107**, (22) 87 (1963).

488. PEARCY, M. A.: Composition Plasma Arc Spraying. Lockheed Aircraft Corp., Sunnydale, Calif., Special Bibliography, No. SB 61-67, 1961.

489. PECHMAN, A.: Ceramics for High-Temperature Applications. Ceram. Age, **62**, (11) 27—31 (1953).

490. PENDLETON, W. W.: Development of Magnet Wires Capable of Operation at 850° C and under Nuclear Radiation. Anaconda Wire and Cable Co., Defense Center, AD-267-176, AD-277-133, 1961—1962.

491. PENDLETON, W. W.: Radiation-Resistant Magnet Wire for Use in Air and Vacuum at 850° C. ASD, Wright-Patterson Air Force Base, TDR 63 164, U.S. Air Force, 1963.

492. PENTECOST, J. L., and H. HAHN: Nonfissionable Ceramic Coatings and Coatings Processes. Joint Conference on Nuclear Applications on Non-Fissionable Ceramics, American Ceramic Society, May, 1966.

493. PETERS, F., and H. ENGELL: Über die Haftfestigkeit von Zünder auf Stahl (On the Adhesive Strength of Oxides on Steel). Arch. F. Eisenhw., **39**, (5) 275—282 (1959).

494. PHELPS, H. C.: Plasma Flame Cutting of Mild Steel Seen as Competitive with Oxy-Fuel Process. Welding Eng., **45**, (12) 33—37 (1960).

495. PHIPPS, G. F., and W. G. SCHREITZ: Mogul Turbo-Jet (Wire Gas) Metallizing Gun and Accessories Submitted by Metallizing Company of America. Marine Engineering Lab., Defense Documentation Center, AD-226 670, 1958.

496. PHIPPS, G. F., and J. H. SIEGEL: Development of Procedure and Application of Cladding to Rotavac Rotor. Defense Documentation Center, AD-94 679 L, 1956.

497. PIROGOV, I. A., R. M. BURON, and A. L. FRIEDBERG: Electrical Properties of Al_2O_3-Nickel Metal Multilayer Flame-Sprayed Coatings. Amer. Ceram. Soc. Bull., **45**, (12) 1071—1074 (1966).

498. PIROGOV, I. A., and L. D. SVIRSKII: Protsessy Formirovanii Pokrytii, Nanosimykh Metodam Gazoplamennogo Napyleniia (Processes of Formation of Coatings Deposited by the Method of Gas-Flame Spraying). Transactions Temperaturo ustorchivye Zashchitnye Pokrytiia Leningrad, 1966.

499. PLANKENHORN, W. J., and D. G. BENNETT: Effect of Ceramic Coatings on the Fatigue Strength of Stainless Steel. Contract AFW 33 (038) dc 14520, U.S. Air Force, 1954.

500. PLUMMER, M.: The Formation of Metastable Aluminas at High-Temperatures. J. Appl. Chem., **8**, 35 (1958).

501. PLUNKETT, J. D.: NASA Contributions to the Technology of Inorganic Coatings. NASA SP-5014, Technology Survey, National Aeronautics and Space Administration (No Date).

502. PODKOVICH, E. G., et al.: The Structure and Properties of Pseudo-Alloys Obtained by Electric and Flame Metal Spraying. TR. Rostovsk, Inst. Sel'sk. Khoz Mashinostroenis, **12**, 46—51 (1959).

503. POE, A. H., and H. E. SHIGLEY: Boeing-Wichita Materials and Research Development Programs, 1957—1961. Defense Documentation Center, AD-271 166, 1961.

504. POREMBKA, S. W., H. D. HANES, and P. J. GRIPSHOVER: Powder Metallurgy of Beryllium. Battelle Memorial Institute, Defense Documentation Center, AD-821 672, 1967.

505. POULOS, N. E., C. A. MURPHY, and J. D. WALTON: Ceramic Systems for Missile Structural Applications. Georgia Institute of Tech., NOW 630143 A 651, 1963.

506. POULOS, N. E., and S. R. ELKINS: High-Temperature Ceramic Structures. Georgia Institute of Technology, NOrd-15701, U.S. Navy, 1962.

507. POULOS, N. E., S. R. ELKINS, and J. D. WALTON: Investigation of High-Temperature Resistant Materials. Georgia Institute of Technology, NOrd-15701, U.S. Navy, 1961.

508. POULSEN, S. C.: The Oxyacetylene Plasmadyne of Plasma Arc. Machine Moderne, **56**, (4) 25—29 (1962).

509. PRATT, D. S., E. SHOFFNER, and E. E. KELLER: Literature Survey. General Dynamics, Defense Documentation Center, AS-402 166, 1959.

510. PRINC, F.: Heißpritzen von Metallen und anderen Stoffen (Hot Spraying of Metals and Other Substances). Tech. Z. f. praktisches Metall., **54**, (1) 27—31 (1960).

511. PUGH, J. W.: Powder Metallurgy of Columbium. General Electric Co., Defense Documentation Center, AD-237 899 (No Date).

512. RAIRDEN, J. R.: Porous Bodies of Tantalum, Niobium, and Aluminum Fabricated by Metal-Spray Processes. General Electric Co., Schenectady, New York, 1966.

513. RANZ, W. E.: On Sprays and Spraying. Pennsylvania State University Engineering Research Bulletin, No. 65, 1965.

514. RAUSCH, J. J., and F. C. HOLTZ: High-Temperature Oxidation Protective Coatings for Vanadium-Base Alloys. Illinois Institute of Technology Defense Documentation Center, AD-296 760, 1963.

515. REAVES, W. A., and E. J. CHAPIN: An Investigation of Barrier Coatings on Graphite Molds for Coating Titanium. Naval Research Lab., Defense Documentation Center, AD-623 992, 1965.

516. REED, B. L., and W. H. JONES: The Investigation of the Nature of the Forces of Adhesion. Defense Documentation Center, (ASTIA) AD-15644 (No Date).

517. REED, L., R. McRAE, and C. BARNES: Ceramic Electron Devices. Eitel-McCullough, Inc., Defense Documentation Center, AD-414 907, 1963.

518. REED, R.: Development of Manufacturing Technology on Organic and Inorganic Foams for High-Temperature Radome Applications. Whittaker Corp., Calif., Defense Documentation Center, AD-824 728, 1967.

519. REININGER, H.: Finishing Supplement Sprayed Metal Coatings. Metal Ind., **52**, (3) 251, (4) 255, 291 (1954).

520. RICHARDSON, L. D.: Ceramics for Aircraft Propulsion Systems. Amer. Ceram. Soc. Bull., **33**, (5) 135—137 (1954).

521. RICHMOND, J. C., H. G. LEFORT, H. WILLIAMS, and W. N. HARRISON: Ceramic Coatings for Nuclear Reactors. J. Amer. Ceram. Soc., **38**, 72—80 (1955).

522. RICHTER, W.: Mischkeramische Werkstoffe für hohe Temperaturen. Silikattech., **8**, (9) 387—389 (1957).

523. RIESEN, A. E.: Tungsten Extrusion Development Program. Wright-Patterson Air Force Base, IR 7 793 V 4, U.S. Air Force, 1962.

524. RILEY, M. W.: New Flame Sprayed Ceramics. Mater. Methods, **42**, (3) 96—98 (1955).

525. ROBIETTE, C. A.: Metal Spraying for Corrosion Protection. Ind. Finishing, **6**, (69) 564 (No Date).

526. RODES, T. W.: Metal Spraying for Materials Handling. Ind. Eng. Chem., **44**, (1) 127 A (1952).

527. ROLAND, E. H., and G. B. HOOD: Development of New and Improved Forging Lubricants. TRW Equipment Labs., Ohio, Defense Documentation Center, AD-822 137 (1967).

528. ROOKSBY, H. P.: The Preparation of Crystalline γ-Alumina. J. Appl. Chem., **8**, (1) 44—49 (1958).

529. ROOT, D. R., and C. H. SAVAGE: Determination of Mechanical and Thermophysical Properties of Coated Refractory and Superalloy Thin Sheet. AFML, Wright-Patterson Air Force Base, RTD-TDR-63-4068, U.S. Air Force, 1963.

530. ROPER, E. H.: An Improved Method of Oxy-Fuel Combustion. Welding J., **34**, (4) 337—344 (1955).

531. ROSE, K.: New Aluminized Coating. Mater. Methods, **43**, (3) 104 (1956).

532. ROSENBERRY, J. W., and J. C. WURST: Test Program for the Evaluation of Protective Coatings for the XLR-99 Thrust Chamber. S.A.M.P.E. National Meeting, November, 1961.

533. ROSENTHAL, J. J., and M. C. TINKLE: Flame Spraying Uranium Oxide. Report L 443961, Los Alamos Scientific Laboratory, New Mexico, 1967.

534. ROTHENBERG, H. C., and G. W. WALTERS: Ferrite Materials for Microwave Frequencies. General Electric Co., Defense Documentation Center, AD-250 183, 1960.

535. ROTREKL, B.: Metal Coated Plastics and Their Application. Army Foreign Science and Technology Center, Defense Documentation Center, AD-457 563, 1965.

536. Rous, W. C.: Selecting Ceramic Coatings for Jet Engine Parts. Mater. Methods, **38**, (6) 116—119 (1953).

537. Rundle, N. L.: Testing and Evaluation of Wear-Resistant Materials for High Performance Hydraulic Pump Components. Paper Presented at 21st National Confernce on Fluid Power, October, 1965.

538. Sagel, K.: Eigenschaften von Spritmetallüberzügen. Metalloberfläche, **11**, 255 (1957).

539. Sales, A. T., et al.: Supersonic Rain Erosion Resistant Coating Materials. Wright-Patterson Air Force Base, AFML-TR-68-364, Part I, Part II, U.S. Air Force, 1968, 1969.

540. Santoli, P. A.: Extruding and Drawing Molybdenum to Complex Thin H-Section. Allegheny Ludlum Steel Corp., Penna., Defense Documentation Center, AD-421 882, 1964.

541. Sauerwald, F.: Über synthetische Festkörper, XIV: Schlagsinterversuche kurzer Dauer (On Synthetic Solids, XIV: Impact Sintering Experiments of Short Duration). Z. f. Physik. Chem., **209**, 206—221 (1958).

542. Sayre, S.: Development of Fused Metal Coatings. Welding J., **31**, (1) 35—39 (1952).

543. Schiller, R. J.: Plasma Arc Fabrication and Coatings. Presented at the Fourth Pacific Area National Meeting of ASTM, Paper 27, October, 1962.

544. Schlichting, H. D., and H. Wulf: Zur Messung statischer Beanspruchungen großer Bauteile bei hohen Temperaturen mit Dehnungsmeßstreifen (Strain-Gauge Measurements of High Static Loads on Large Structural Components at High Temperature). Archiv für technisches Messen und industrielle Meßtechnik. 1967.

545. Schoop, M. V.: A Modern Electric Arc Pistol. Electroplat. Metal Spray., (6) 33 (1964).

546. Schoop, M. U., and H. Guenther: Das Schoopsche Metallspritzverfahren (The Schoop Metal Spraying Technique). Stuttgart: Franckh Publishers, 1917.

547. Schulz, D., P. Higgs, and J. Cannon: Research and Development on Advanced Graphite Materials Vol. XXXIV Oxidation—Resistant Coatings for Graphite. Wright-Patterson Air Force Base, WADD TR-61-72, U.S. Air Force, 1964.

548. Schwartz, H.: Plastics for Flame Spraying and Their Characteristics. Plastics and Rubber, **4129**, 13—26 (1965).

549. Seghezzi, H., and E. Gebhardt: Aufbau und Bewertung von Flammspritzschichten (Formation and Evaluation of Flame Sprayed Layers). Ind.-Anz., **80**, (49) 708—713 (1958).

550. Selover, T. B.: Properties of Nickel Fume Generated in a Plasma Jet. Metal Progr., **82**, (8) 1544 (1962).

551. Shepard, A. P.: Sealer Extends Rust-Free Life for Metallized Structures. Iron Age, **191**, (15) 69—71 (1963).

552. Sherwood, P. W.: Ultra-High Temperature Spraying. Prod. Finishing, **14**, (11) 98—100 (1961).

553. Shkliarevskii, E. E., and E. V. Smirnov: Udel'Noi Elektricheskoe Sprotivlenie Okisi Aliuminiia Nanesennoi Na Podlozhku Plazmennym I Gazoplamennym Napyleniem (Electrical Resistivity of Aluminum Oxide Deposited on a Substrate by Plasma and Gas-Flame Spraying). Inzhenerno-Fizicheskii Zhurnal, **6**, 713—716 (1969).

554. Shlyakova, K. S.: The Structure and Certain Properties of Gas-Flame Coatings. Translated by Joint Public Research Service, Report No. 16788, 1962.

555. Shoffner, J. E., E. E. Keller, and W. M. Sutherland: Materials Finishes and Coatings—Flame Sprayed—Alumina, Zirconia Tungsten Carbide. Rocket Blast Impingement Resistant. Report No. 8926-060, General Dynamics/Convair, 1958.

556. Simpson, H. G.: Evaluation of Protective Finishes for Highly Susceptible Stress Corrosion Areas. Lockhead-Georgia Co., Defense Documentation Center, AD-840 050 L, 1967.

557. SINGLETON, R. H., E. L. BOLIN, and F. W. CARL: The Fabrication of Complicated Refractory Metal Shapes by Arc Spray Techniques. High-Temperature Materials, Part 2, Metallurgical Society Conferences, **18**, 641—654, Technical Conference Proceedings. Ed. by G. M. AULT, W. F. BARCLAY, H. P. MUNGER, New York: Inter-Science Publishers Div. John Wiley and Sons, Inc., 1963.

558. SINGLETON, R. H., E. L. BOLIN, and F. W. CARL: Tungsten Fabrication by Arc Spraying. J. Metals, **13**, (7) 483—486 (1961).

559. SINGLETON, R. H., *et al.*: The Fabrication of Complicated Refractory Metal Shapes by Arc Spray Techniques. Western Metals Congress, March, 1961.

560. SITA, E. R.: Development of the Flame Sprayed Aluminum Concept for Conductive Cooling of XE Engine-Mounted Electrical Cables. NASA-CR-104090, National Aeronautics and Space Administration, 1967.

561. SITZER, D. H.: Flame Sprayed Nickel Aluminide Coatings. Metal Progr., **86**, (9) 128 (1964).

562. SKLAREW, S.: Marquardt Process Specification Number 512, Application of Rokide Coatings. Marquardt Aircraft Co., 1956.

563. SKLAREW, S.: Emittance Studies of Various High-Temperature Materials and Coatings. Marquardt Corp., Defense Documentation Center, AD-299 417, 1963.

564. SMITH, D. M.: Plasma Spraying of Refractory Materials. Gen. Motors Eng. J., 2nd Quarter, 1963.

565. SMITH, H. E., and J. C. WURST: The Evaluation of High-Temperature Materials Systems with an Arc-Plasma-Jet. Wright-Patterson Air Force Base, TDR-64-73, U.S. Air Force, 1964.

566. SMITH, H. E.: Development and Investigation of an Arc-Plasma Material Evaluation Facility. Wright-Patterson Air Force Base, ASD-TDR-62-653, U.S. Air Force (No Date).

567. SMITH, R. S., N. STEPHENSON, and T. A. TAYLOR: The Influence of Some Process Variables on the Adhesion of Aluminum Spray Coatings. National Gas Turbine Establishment Pyestock, England, Defense Documentation Center, AD-233 639, 1959.

568. SMITH, R. S., and N. STEPHENSON: An Adhesion Test for Aluminum Spray-Coatings and Other Metallised Surfaces. National Gas Turbine Establishment, Pyestock, England, Defense Documentation Center, AD-233 638, 1959.

569. SMITH, R. S., and N. STEPHENSON: The Response of Aluminum-Spray Coatings to Simulated Abnormal Service Variables. National Gas Turbine Establishment, Pyestock, England, Defense Documentation Center, AD-236 531, 1960.

570. SMOKE, E. J., and C. J. PHILLIPS: Inorganic Dielectrics Research. New Jersey Ceramic Research Station, Rutgers, Defense Documentation Center, AD-287 173, 1962, AD-287 167, 1960, AD-287 171, 1961.

571. SMOKE, E. J., and D. R. ULRICH: Devitrified Barium Titanate Dielectrics. Amer. Ceram. Soc. J., **49**, (4) 210—215 (1966).

572. SPACHNER, S. A.: Liner for Extrusion Billet Containers. ASD, Wright-Patterson Air Force Base, TR-7-945-Pl, U.S. Air Force, 1962.

573. SPENCER, D. J.: High Mach Number and Materials Research Program Phase II, Arc Plasma Investigations and Arc Tunnel Materials Studies. Wright-Patterson Air Force Base, TDR-594, U.S. Air Force, 1961.

574. SPITZIG, W. A.: Sintering of Arc Plasma Sprayed Tungsten. M.S. Thesis, Case Institute Tech., 1962.

575. STACKHOUSE, R. D., *et al.*: Plasma-Arc Plating. Prod. Eng., **29**, (50) 104—106 (1958).

576. STARR, G. A.: Hydrofoil Materials Research Program. Ling-Temco-Vought, Inc., Defense Documentation Center, AD-464 356, 1964.

577. STEFFENS, H. D.: Der Lichtbogen-Metallspritzprozeß (The Arc Metal Spraying Process). Jahrbuch der Oberflächentechnik, 205—215, Berlin: Metall-Verlag, 1963.

578. STEFFENS, H. D.: The Bonding Mechanism in Metal Spraying. Report given at the Third International Metal Spraying Conf., Madrid, 1962.

579. STEFFENS, H., and M. DITTRICH: Molybdän als Spritzwerkstoff (Molybdenum as Spray Materials). Schw. U. Schn., **15**, (3) 97—106 (1963).

580. STERRY, W. M.: Ceramic and Composite Ceramic-Metal Materials Systems Applicable to Re-entry Structures. Paper presented at the Society of Aircraft and Materials and Process Engineers Symposium, Dayton, Ohio, March, 1960.

581. STETSON, A. R., and C. A. HAUCK: Plasma-Spraying Techniques for Toxic and Oxidizable Materials. J. Metals, **13**, (7) 479—482 (1961).

582. STOKES, C. S., and W. W. KNIPE: The Plasma Jet in Chemicals Synthesis. Ind. Eng. Chem., **52**, 287—288 (1960.)

583. STRAUSS, E., and S. SEIGLE: Development of Composite Rocket Nozzles. TR-61-109-B, NOW-61-0479-C, U.S. Navy (No Date).

584. STUBBS, V. R.: Development of a Thermal Barrier Coating for Use on a Water-Loaded Nozzle of a Solid Propellant Rocket Motor. NASA-CR-72549, National Aeronautics and Space Administration, 1969.

585. STRUBLER, G. L.: Means for Applying Vitreous Enamel. U.S. Patent 2,085,278, 1937.

586. STRUBLER, G. L.: Method of Coating Articles with Glass. U.S. Patent 2,356,016, 1944.

587. STUDT, T. B., A. E. KING, A. J. KRAUSE, and W. J. SHILLING: 300 C Rotating Rectifier Alternator, Phase II—Design, Fabrication and Test of Alternator. Westinghouse Electric Corp., Ohio, Defense Documentation Center, AD-853-454, 1969.

588. SULLY, A. H.: Special Coatings for Metals Used at High-Temperatures. Prod. Eng., **25**, (8) 135—141 (1954).

589. SUMNER, E. V.: Study for AM-2 Mat Surfacing. Harvey Engineering Lab., Calif., Defense Documentation Center, AD-842-640-L, 1967.

590. SUNNEN, J.: An Arc Device for the Application of Very High-Temperatures. Arcos, **38**, 47—54 (1961).

591. SYCHROVSKY, H.: Entwicklung der Gasturbinenwerkstoffe. M.T.Z., **16**, 202—205 (1955).

592. TETER, M. A.: Flame Plated Coatings. Mater. Methods., **43**, (2) 100—102 (1956).

593. THOMPSON, V. S.: Structural Changes on Reheating Plasma-Sprayed Alumina. Amer. Ceram. Soc. Bull., **47**, (7) 637—674 (1968).

594. THOMPSON, V. S.: Aluminum Flame Sprayed Coating Process for Reinforced Plastic Aircraft Assemblies. Welding J., **47**, (1) 31—36 (1968).

595. THOMPSON, V. S.: Evaluation of Flame Spray Mandrel Materials. MDR 2 24775 Boeing Co., Defense Documentation Center, AD-454 22, 1964.

596. THOMPSON, V. S.: Properties of Stainless Steel and Alumina Flame-Sprayed Free-Standing Shapes. MDR 2-24773, Boeing Co., Defense Documentation Center, AD-468 900, 1965.

597. THORPE, M. L.: Plasma Flame Technology Handbook. Thermal Dynamic Corp., Bulletin 132, 1959.

598. THORPE, M. L.: Plasma Arc Equipment. French Patent 1,260,262, 1960.

599. THORPE, M. L.: The Plasma Jet and its Uses. Res. Develop., **11**, (1) 10 (1960).

600. TINER, N. A.: Refractory Coatings for Aerospace Applications. Amer. Inst. of Chem. Eng. Annual Meeting, Symposium on Aerospace Materials No. 52, December, 1963.

601. TOLOTTA, S., and J. J. AGOZZINO: Development of a Flame-Sprayed Coating Technique for the Installation of Instrumentation on Naval Machinery Components. Naval Ship Engineering, Defense Documentation Center, AD-830-645-L, 1968.

602. TOTTEN, J. K.: Ramjet Technology Program 1963, Section XVIII High-Temperature Coated Tungsten Structures. Marquardt Corp., Defense Documentation Center, AD-602 049, 1964.

603. TOUR, S.: Modern Developments in Metallizing. Welding J., **31** (3) 109 (1952).

604. TOUR, S.: Fused in Place Spray Metallized Coatings. Welding J., **34**, (4) 329 (1955).
605. TRAVITSKAIA, E. O.: Temperaturoustoichivye Zashchitnye Pokrytiia (Temperature-Stable Protective Coatings). Seminar Po Zharostoikim Pokrytiiam, 3rd Leningrad, U.S.S.R., 1966, Trudy, Izdatel Stvo Nauka, 1968.
606. TRIAS, J., F. M. GRABER, and E. E. KELLER: Materials-Finishes and Coatings-Aluminum Foil, Flame Sprayed Aluminum and Flame Sprayed Tin Reflecting Surfaces. Reflectant Characteristics. Report No. 8926-105, General Dynamics, Convair, Contract AF 33 (657)-8926, U.S. Air Force, 1960.
607. TRIPP, H. P.: Development of Corrosion Resistant Coatings for Use at High-Temperatures. Gulton Ind., New Jersey, Defense Documentation Center, AD-691 850, 1960.
608. TROUT, O. F.: Exploratory Investigation of Several Coated and Uncoated Metal, Refractory and Graphite Models in a 3,800° F Stagnation Temperature Air Jet. National Aeronautics Space Administration, Defense Documentation Center, AD-231 523 (No Date).
609. TYLER, P. M.: Plasma for Extractive Metallurgy. J. Met., **13**, (1) 51—54 (1961).
610. UMBAUGH, C. W.: Development of Thin Organic Rolled Film Capacitor. Linde Div., Union Carbide Corp., Defense Documentation Center, AD-615 753, 1954.
611. UNGER, R.: Arc Sprayed Gradated Coatings as Applied to the X-15 Thrust Chamber. S.A.M.P.E., National Meeting, November, 1961.
612. UNGER, R., J. W. ROSENBERRY, and J. C. WURST: Arc Sprayed Graduated Coatings as Applied to the X-15 Thrust Chamber, Part I, Application, Part II Testing. Presented to the S.A.M.P.E. Symposium on Ceramic and Composite Coatings and Solid Bodies, November, 1961.
613. UNGER, R.: Composite Refractory Coatings. West Coast Regional Meeting, Amer. Ceram. Soc., October, 1960.
614. UNGER, R.: Investigation of Nickel-Aluminum Oxide Coatings on Atlas Sustainer Chamber Serial No. ROO 3 R Type 200860. Rocketdyne Report CEM 01114-504, January, 1960.
615. UNGER, R.: Testing of Ceramic Coated B-3 Type (Thor) Chamber, Serial No. 142. Rocketdyne Report CEM 914-677, August, 1959.
616. VAGI, J. J., et al.: Review of Recent Developments in Metals Joining. Defense Metals Information Center, Battelle Memorial Institute DMIC Memo., 125, 1961.
617. VANDERPOOL, H., and A. SHEPARD: Metallizing, Chapt. 10, Surface Protection Against Wear and Corrosion. Amer. Soc. Met., Metals Park, Ohio, 1954.
618. VASILOS, T., and G. HARRIS: Impervious Flame-Sprayed Ceramic Coatings. Amer. Ceram. Soc. Bull., **41**, (1) 14—17 (1962).
619. VENEZKY, D. L., A. G. SANDS, and E. B. SIMMONS: Protective Coatings for Magnesium Alloys, Part I Effect on Mechanical Properties of a New Technique for Fusing Teflon to Magnesium and Aluminum Alloys. Naval Research Lab., -6209, U.S. Navy, 1965.
620. VENEZKY, D. L., A. G. SANDS, and E. B. SIMMONS: Protective Coatings for Magnesium and Aluminum Alloys to Corrosion by 3% Sodium Chloride Solution. Naval Research Lab., -6353, U.S. Navy, 1965.
621. VIDANOFF, R. B.: Silica Fiber Forming and Core-Sheath Composite Fiber Development. Whittaker Corp., Calif., Defense Documentation Center, AD-410 051, 1963.
622. VIGLIONE, J.: Effect of Plasma Sprayed Metco 404 and 439 Coatings on the Fatigue Properties of 4340 Steel. Naval Air Development Center, MA-6923, U.S. Navy, 1969.
623. VOGAN, J. W., and J. L. TRUMBULL: Metal-Ceramic Structural Composite Materials. Wright-Patterson Air Force Base, TDR-64-83, U.S. Air Force, 1964.
624. VON HOFE, A.: Oberflächenbeschaffenheit und Haftfestigkeit von Flammspritzschichten (Surface Condition and Adhesive Strength of Flame Sprayed Layers). Ind.-Anz., **79**, (37) 542—543 (1957).

625. VON WHARTENBERG, H.: Zur Kenntnis der Tonerde. Z. Anorg. Allg. Chem., **269**, 76—85 (1952), **270**, 328 (1952).

626. VOSSEN, J. L., and R. E. WHITMORE: Plasma Anodized Lanthanium Titanite Films Used as Evaporable Thin Film Dielectric, Investigating Electric Properties. IEEE Trans. on Parts, Materials and Packaging, **PMP-1** (6) 105—155 (1965).

627. WADE, W. R., and W. S. SLEMP: Measurements of Total Emittance of Several Refractory Oxides, Cermets and Ceramics for Temperatures from 600° F to 2000° F. NASA TN D-998, National Aeronautics and Space Administration, 1962.

628. WALKER, D. L., and R. S. KIRBY: Process Variables in Plasma Spraying, Refractory Coatings on Beryllium. Lockheed Aircraft Corp., ATL-574, Contract NOrd 17017, U.S. Navy, 1967.

629. WALKER, D. R., and H. W. WYATT: Measuring Fracture of Ceramic Coatings. Mater. Design Eng., **48**, (2) (1958).

630. WALTON, J. D.: Experimental Application of Arc and Flame Sprayed Coatings. Technical Paper Presented at Aerospace Finishing Symposium Fort Worth, Texas, December, 1959.

631. WALTON, J. D., J. D. FLEMING, and N. E. POULOS: Investigation of High-Temperature Resistant Materials. Georgia Inst. of Tech. Defense Documentation Center, AD-222 920 (No Date).

632. WALZ, F. C.: Development of Film Type 2000 Degrees F Heaters. Wright-Patterson Air Force Base, ASD-TDR-62-662, U.S. Air Force, 1964.

633. WATKINS, L. L.: Corrosion and Protection of Steel Piling in Seawater. Army Coastal Engineering Research Center, Defense Documentation Center, AD-690 803, 1969.

634. WATSON, D. A.: Sprayed Metal Coatings in Product Design. Mater. Methods, **42**, (6) 106—109 (1955).

635. WATTS, A. A.: Development of High Strength Materials for Solid Rocket Motors. General Electric Co., Defense Documentation Center, AD-236 126, 1960.

636. WAXMAN, A. S., and K. H. ZAININGER: Radiation Resistance of Al_2O_3 MOS Devices. IEEE Transactions on Electron Devices, **ED-16**, (4) 333—338 (1969).

637. WAUGH, J. S., A. E. PALADINO, J. J. GREEN, and W. R. BEKEBREDE: Fine-Grain Dense Ferrites. Raytheon, Mass., Defense Documentation Center, AD-403 292, 1963.

638. WESTBROOK, A. H.: Metal Ceramic Composites. Amer. Ceram. Soc. Bull., **31**, (6) 205 to 208, 248—250 (1952).

639. WESTERHOLM, R. J., J. C. McGEARY, R. L. WOLFF, and H. M. HUFF: Flame Sprayed Coatings. Machine Design., **33**, (18) 82—92 (1961).

640. WEYMOUTH, L. J.: Strain Gage Application by Flame Techniques. Instrument Society of America, 19th Annual Conf., New York, October, 1955.

641. WHEILDON, W. M.: Coating Metals and Other Materials with Oxide and Articles Made Thereby. U.S. Patent 2, 707, 691, 1955.

642. WHEILDON, W. M.: The Role of Flame Sprayed Ceramic Coatings as Materials for Space Technology and the Systems for Application. Presented at A.I.Ch. E. Meeting New Orleans, La., March, 1963.

643. WHEILDON, W. M.: Flame Sprayed Ceramic Coating, Survey and State of the Art. Presented at the Amer. Soc. for Testing Mater. June, 1966.

644. WHEILDON, W. M.: Oxide Coatings-Resistant to Temperatures in Excess of 3000° F and to Erosion of Supersonic Velocities. Pacific Coast Cer. News, **4**, (11) 21—25 (1955).

645. WHEILDON, W. M.: Wear Resistant Ceramic Tooling in the Fabrication of Ceramic Shapes. Amer. Ceram. Soc. Bull., **42**, (5) 308—311 (1963).

646. WHEILDON, W. M.: Flame-Sprayed Ceramic Coatings in Space Technology. Spaceflight, **9**, (2) 55—62 (1967).

647. WHITTAKER, J. A.: Corrosion Fatigue of High-Strength Aluminum Alloys as Affected by Notches, Anodic Metal Coatings, and Applied Cathodic Currents. Fulmer Research Institute Ltd., Defense Documentation Center, AD-415 264, 1963.

648. WHITTAKER, J. A.: Study of the Exfoliation Corrosion of Aluminum Alloys. Fulmer Research Institute Ltd., Defense Documentation Center, AD-245 779, 1960.

649. WHYMARK, R. R., and W. E. LAWRIE: Ultrasonics and Ceramic Coatings. Wright-Patterson Air Force Base, ASD-TR-60-157, U.S. Air Force, 1960.

650. WIGG, L. D.: The Effect of Scale on Fine Sprays Produced by Large Air-Blast Atomizers. National Gas Turbine Establishment, Defense Documentation Center, AD-227 044, 1959.

651. WILSON, J. W.: Refractory Materials a Special Projects Office. Stanford Research Inst., Calif., Defense Documentation Center, AD-233 955, 1960.

652. WILSON, R. E.: Ceramic-Plastic Composites for Rain Erosion Resistant Radomes. Naval Ordnance Lab., NOLTR-65-63, U.S. Navy, 1965.

653. WINTERMUTE, G.: Development of Manufacturing Techniques for Production of Rain Erosion Resistant Coated Structures. Goodyear Aerospace Corp., Arizona, Defense Documentation Center, AD-823 303, 1967.

654. WLASSOW, A., and K. SWINKOW: High Frequency Metal Spraying. Moscow: Maschgia Publishers, 1960.

655. WLODEK, S. T.: Coatings for Columbium. J. Electrochem. Soc., **108**, 177 (1961).

656. WOOD, R. A.: Surface Treatment of Titanium. Defense Metals Information Center, Defense Documentation Center, AD-463 016, 1965.

657. WURST, J. C.: The Evaluation of High-Temperature Materials, Part I: Evaluation of Coatings for Refractory Alloys, Part II: Materials Evaluation with an Arc-Plasma-Jet. Wright-Patterson Air Force Base, TDR-64-62, U.S. Air Force, 1964.

658. WURST, J. C., J. A. CHERRY, D. A. GERDEMAN, and N. L. HECHT: The Evaluation of Materials Systems for High-Temperature Aerospace Applications. Wright-Patterson Air Force Base, AFML TR-65-339, Part I, U.S. Air Force, 1966.

659. WURST, J. C., and J. A. CHERRY: A Quarterly Progress Report on the Evaluation of High-Temperature Materials. University of Dayton, Contract AF 33 (616)-7838, U.S. Air Force, 1963.

660. WURST, J., J. CHERRY, and N. L. HECHT: The Evaluation of High-Temperature Materials. Research Institute University of Dayton, Defense Documentation Center, AD-478 564, 1964.

661. YATES, J.: Materials Property Data. Bendix Products Div. Task 73812, Contract AF 33(616)-8086, U.S. Air Force, 1961.

662. YEZERSKIY, A. N.: Flame-Spraying of Plastics. Foreign Tech. Div., Wright-Patterson Air Force Base, Defense Documentation Center, AD-405 250 (No Date).

663. ZANDSTRA, K. A.: Progress Review No. 58: Plasma Heating. J. Inst. Fuel, **38**, (297) 450—455 (1965).

664. ZILBERBERG, V. G.: Prochnost Stsepleniia Plazmennykh Pokrytii S Osnovoi. Poroshko-vaia Met., **9**, 96—100 (1969).

665. ZOLTOWSKI, P.: Les Couches En Alumine Effectuées Au Pistolet A Plasma. Parte I, Rev. Inst. Hautes Temper et Refract. **t5**, 253—265 (1968), Parte II, Rev. Ind. Hautes Temper et Refract. **t6**, 65—70 (1969).

666. First Refractory Composite Working Group Meeting (1958) compiled by:
ROLLER, D.: Summary of the First High-Temperature Inorganic Refractory Coatings Meeting. Wright Air Development Center, WCLT-TM-58-139, U.S. Air Force, 1958.

667. Second Refractory Composite Working Group Meeting (1959) compiled by:
ROLLER, D.: Summary of Second High-Temperature Inorganic Refractory Coatings Working Group Meeting. WADT-TR-69-415 Wright Air Development Center, U.S. Air Force, 1959.

a. WALTON, J. D.: Arc Sprayed Coatings; pp. 7—8.

b. BRADSTREET, S. W.: Effects of Particle Chill Rate; p. 11.

c. INGHAM, H. S.: Evaluation Methods and Equipment for Flame Sprayed Coatings; pp. 15—28.

d. THORPE, M. L.: Plasma Jet Spraying; pp. 29—36.

e. HERRON, R. H.: Research at Bendix Aviation Corp.; pp. 41—43.

f. WHEILDON, W. M.: Rokide Flame Spraying; pp. 51—54.

g. GRINTHAL, R.: Rod Flame Spraying; pp. 58—59.

h. SKLAREW, S.: Flame Sprayed Ceramics; pp. 62—66.

i. LEVINSTEIN, M.: Columbium and Columbium Alloys; pp. 67—69.

j. SINGLETON, R. H.: Refractory Coating Research; pp. 71—74.

k. PENTECOST, J. L.: Coating Research; pp. 91—93.

l. OLCOTT, E. L.: Coatings for Rocket Motors; pp. 105—106.

m. LEEDS, D. H.: Flame Sprayed Coatings; pp. 107—108.

n. MOORE, D. G.: Flame Spraying; pp. 109—110.

668. The Fourth Composites Working Group Meeting, November, 1960, Preprints:

a. SMITH, G. D.: Heat Transfer Characteristics of Materials in Plasma Flames.

b. MASON, C. R., and J. D. WALTON: Thin Arc Sprayed Shapes II.

c. EISENLOHR, A.: The Role of Materials in Plasma Flame Spraying.

d. STERRY, W. M.: A Progress Report on Boeing Studies of Refractory Coatings, Metal-Ceramic Composites for Thermal Protection Systems.

e. WINTERSTEEN, R.: Sprayed Deposition with a Plasma-Therm.

f. WEARE, N. E.: Refractory Coating Research at Advanced Technology Laboratories.

g. SINGLETON, R. H.: Fabrication of Tungsten Shapes Using the Arc Plasma Spray Forming Technology.

h. BURNETT, P. L.: AVCO Plasma Spray Program.

i. LEVINSTEIN, M. A., and W. B. HALL: Properties of Plasma Materials.

j. GIANNINI: Plasmadyne Corp., A Survey: Contributions for the Establishment of Standard Materials Screening Tests for High-Temperature Materials Section.

k. THORPE, M. L.: Progress Report Plasma Generators and Associated Equipment.

l. TROSTEL, L. J.: Rokide Coating Development.

669. The Fifth Refractory Composite Working Group Meeting (1961), compiled by: HJELM, L. N.: in Summary of the Fifth Refractory Composites Working Grouping Meeting. Air Force Systems Command, ASD-TDR-63-96, U.S. Air Force, 1961.

a. UNGER, R.: Composite Plasma Depositions; No. 19.

b. MASH, D. R.: Refractory Coatings Research at Advanced Technology Laboratories; No. 25.

c. LEEDS, D. H.: Interface Bonding Studies and a New Plasma Arc Spray Gun, Accomplishment and Plans; No. 26.

d. ROSENBERRY, J. W., and J. C. WURST: High-Temperature Materials Evaluation; No. 27.

e. DAVIS, L. W.: Activities in the High-Temperature Inorganic Refractory Coating Field; No. 28.

f. MURPHY, C. A., N. E. POULOS, and J. D. WALTON: Thin Arc-Sprayed Shapes III; No. 30.

g. GRISAFFE, S. J.: A Preliminary Investigation of the Thermal Heating of Powder Particles During Plasma Spraying; No. 31.

h. JOHNSON, R. L., and W. M. WHEILDON: Development and Evaluation of a Rokide Plasma System; No. 32.

i. HILL, V. L.: Transition and Temperature and Flexural Strength of Tungsten Materials, Fabrication and Properties of Tungsten; No. 33.

  j. LEVY, M.: Refractory Coating Research and Development Watertown Arsenal Laboratories; No. 35.

  k. BLITON, J. L., and S. W. BRADSTREET: Pertinent Activities in Refractory Composites; No. 36.

  l. JOHNSON, R. L.: Emittance Coatings for the Enhancement of Radiation Heat Transfer from Solids; No. 37.

670. The Sixth Refractory Composite Working Group Meeting (1962), compiled by:
  HJELM, L. N.: in Summary of the Sixth Refractory Composites Working Group Meeting. Volume I and II, Air Force Systems Command, ASD-TDR-63-610, U.S. Air Force, 1962.

  a. STETSON, A. R.: Tungsten Wire Reinforces Plasma Arc Sprayed Tungsten; pp. 187—196.

  b. BLITON, J. L.: Refractory Composites and Coatings; pp. 226—247.

  c. BARTEL, E. H.: Aircraft Coatings for Rocket Blast; pp. 669—680.

  d. AULT, N. N.: Optical Properties for Rokide Coatings; pp. 684—698.

  e. MOLELLA, D. J.: Evaluation of Oxidation Resistant Coatings in a Water-Stabilized Electric Arc at Temperatures to 2325° C; pp. 749—777.

  f. LEEDS, D. H.: Summary Observation of the Interface Between Plasma Sprayed Tungsten and 1020 Steel; pp. 778—787.

  g. LOHRIE, B.: Economic Aspects of Arc Plasma Processing; pp. 788—793.

  h. SMITH, B. D.: Microstructure of Selected Sprayed Coatings on Unalloyed Columbium; pp. 794—800.

  i. GRISAFFE, S. J.: Practical Plasma Spraying; pp. 801—805.

  j. INGHAM, H. S.: Evaluation Methods for Flame Spray Coatings; pp. 806—827.

  k. UNGER, R.: Plasma Fusion Surfacing; pp. 828—840.

  l. JENSON, G. A., and P. L. BURNETT: Some Effects of Tungsten Powders on Plasma Sprayed and Sintered Structure; pp. 841—850.

671. The Seventh Refractory Composite Working Group Meeting (1963), compiled by:
  HJELM, L. N., and D. R. JAMES: in Summary of the Seventh Refractory Composites Working Group Meeting. Volumes I, II, III, Air Force Systems Command, RTD-TDR-63-4131, U.S. Air Force, 1963.

  a. BLITON, J. L.: Plasma-Sprayed Oxide and Vapor-Deposited Nitride Coatings on Tungsten as a Means of Achieving Oxidation Protection; pp. 203—208.

  b. GRISAFFE, S. J., and W. A. SPITZIG: Observations on Metallurgical Bonding Between Plasma Sprayed Tungsten and Hot Tungsten Substrates; pp. 209—226.

  c. LEEDS, D. H.: A Portfolio of Experience in Refractory Metal Protective Systems Bibliography; pp. 442—449.

  d. LEVY, L.: Refractory Coating Research and Developments at U.S. Army Materials Research Agency; pp. 450—467.

  e. BRAMER, S. E.: Plasma Arc Sprayed Free Standing Shapes for Radiation-Cooled Thrust Chambers; pp. 607—623.

  f. EISENLOHR, A.: Arc Plasma Sprayed Tungsten as an Engineering Materials; pp. 632—638.

  g. HEADMAN, W.: Report to Refractory Composites Working Group; pp. 763—781.

  h. TINER, N. A.: Plasma Arc Spraying For Use in Radiating Rocket Chamber; pp. 782—802.

  i. INGHAM, H. S., and F. N. LONGO: New Technique for Plasma Spraying Tungsten; pp. 803—810.

  j. LUSHER, G. H., and W. M. WHEILDON: Development of Heat Generating Systems for Rokide Flame-Spraying; pp. 908—915.

672. Eighth Refractory Composites Working Group Meeting (1964), compiled by:
  JAMES, D. R., and L. N. HJELM: in Summary of the Eighth Refractory Composites

Working Group Meeting. Volumes I, II, III, Air Force System Command, ML-TDR-64-233, U.S. Air Force, 1964.

    a. BLITON, J. L.: Refractory Composite and Coating Work Conducted at IIT Research Institute; pp. 660—679.

    b. TINER, N. A.: Refractory Coatings by Plasma Spraying for Aerospace Applications; pp. 674—707.

    c. DAVIS, L. W.: Activities in the High-Temperature Inorganic Refractory Coatings Field; pp. 708—715.

    d. BRUCKART, W. L., and S. E. BRAMER: Plasma Arc Sprayed Thrust Chamber Development and Testing; pp. 716—766.

    e. GRISAFFE, S. J., and W. A. SPITZIG: On Plasma Sprayed Diffusion Bonding Between Tungsten and Cool Stainless Steel Substrates; pp. 767—774.

    f. WHEILDON, W. M.: Study of Stabilized Methylacetylene-Propadiene as a Fuel Gas for the Rokide Process; pp. 775—785.

    g. HEADMAN, M. L.: Progress Report on Plasma Sprayed Refractory Materials; pp. 786—797.

    h. HARRIS, G. M., and P. L. BURNETT: Recent Developments at AVCO RAD in the Refractory Composites Field; pp. 798—808.

    i. ROSENBERRY, J. W.: Recent Progress in Plasma Arc Spraying and Environmental Simulation at Giannini Scientific Corp.; pp. 809—834.

673. Ninth Refractory Composite Working Group Meeting (1965), compiled by:
HJELM, L. N., D. JAMES, and E. BEARDSLEY: in Summary of the Ninth Refractory Composites Working Group Meeting. Air Force Systems Command, AFML-TR-64-398, U.S. Air Force, 1965.

    a. BEAVER, W. W., and R. M. PAINE: Summary of Activities in the Area of Refractory Composites; pp. 327—352.

    b. UNGER, R.: Plasma Jet Operations; pp. 617—639.

    c. BORTZ, S. A.: Four Short Review Sections from Two Current Major Programs at IIT RI; pp. 825—861.

    d. BROWNING, M. E., et al.: Activities Report on High-Temperature Materials and Coatings Work at AMF; pp. 938—967.

    e. INGHAM, H. S.: Some Advances in Combustion and Plasma Flame Spray Coatings; pp. 968—977.

674. Tenth Refractory Composite Working Group Meeting (1965), compiled by:
HJELM, L. N., D. R. JAMES, and E. H. BEARDSLEY: in Summary of the Tenth Refractory Working Group Meeting. Air Force Systems Command, AFML-TR-65-207, U.S. Air Force, 1965.

    a. INGHAM, H. S., and F. N. LONGO: Flame Sprayed Molybdenum; pp. 161—169.

    b. GRISAFFE, S. J.: Shear Bond Strength of Aluminum Coatings; pp. 336—341.

    c. WHEILDON, W. M.: Status of Rokide Coating System and Summary of Hot Pressing Capabilities; pp. 342—353.

    d. BORTZ, S. A.: A Review of Current Refractory Composite Research in Ceramics at IIT RI; pp. 774—817.

675. Eleventh Refractory Composite Working Group Meeting (1966), compiled by:
JAMES, D. R., and E. H. BEARDSLEY: in Summary of the Eleventh Refractory Composites Working Group Meeting. Air Force Systems Command, AFML-TR-66-179, U.S. Air Force, 1966.

    a. CHAO, P. J.: Composite Materials Development at Wright-Aeronautical Division; pp. 475—487.

    b. BORTZ, S. A.: A Review of Current Refractory Composite Research in Ceramics; pp. 1119—1156.

676. Twelfth Refractory Composite Working Group Meeting (1967), compiled by:
JAMES, D. R., and E. H. BEARDSLEY: in Summary of the Twelfth Refractory Composites Working Group Meeting. Air Force Systems Command, AFML-TR-67-228, U.S. Air Force, 1967.
   a. WHEILDON, W. M.: An Experimental Abrasion Test System and Summary of Results; pp. 385—390.
   b. INGHAM, H. S.: A New Bond Test for Coatings; pp. 391—403.
677. Thirteenth Refractory Composites Working Group Meeting (1968), compiled by:
BEARDSLEY, E. H.: in Summary of the Thirteenth Refractory Composites Working Group Meeting. Air Force Systems Command, AFML-TR-68-84, U.S. Air Force, 1968.
   a. KENSHOL, K. W.: High-Temperature Protective Coatings on Extrusion Dies; pp. 346—352.
   b. WHEILDON, W. M.: Properties of Thermal Sprayed Zirconate Coatings; pp. 353—361.
   c. CARPENTER, H. W.: Heat Barrier for Advanced Rocket Engines and High-Temperature Insulation Systems; pp. 390—386.
   d. WURST, J. C.: Refractory Composite Evaluation at the University of Dayton Research Institute; pp. 362—379.
678. Fourteenth Refractory Composites Working Group Meeting (1968), compiled by:
PURCELL, G. V., and J. P. BOYER: in 14th Refractory Composites Working Group Meeting. Air Force Systems Command, AFML-TR-68-129, U.S. Air Force, 1968.
679. Fifteenth Refractory Composites Working Group Meeting (1969), compiled by:
MAYKUTH, D. J., and K. R. HANBY: in Summary of the Fifteenth Refractory Composites Working Group Meeting. DMIC Report S-26, Defense Metals Information Center, 1969.
   a. BUCKLEY, J. D.: Preliminary Investigation of the Thermal Stress and Insulation Properties of Stabilized Zirconia and Cermet Coatings on Thin-Gage Refractory Alloy Sheet Metal; pp. 5—6.
   b. HEESTAND, R. L., et al.: Recent Developments in Plasma Spray Studies at Battelle-Columbus; pp. 15—16.
680. Sixteenth Refractory Composite Working Group Meeting (1969), compiled by:
HANBY, K. R., and D. J. MAYKUTH: in Summary of the Sixteenth Refractory Composites Working Group Meeting. DMIC Report S-30, Defense Metals Information Center, 1969.
   a. DITTRICH, F. J., and H. S. INGHAM: Thermal Shock Resistance of Plasma Sprayed Zirconia Coatings; p. 1.
681. Seventeenth Refractory Composites Working Group Meeting, June, 1970, Preprints:
   a. HECHT, N. L.: Refractory Coatings Development at University of Dayton Research Institute.
   b. RANSONE, P. O., and I. O. MACCONOCHIE: Summary Review of Integrated Thermal Protection Systems for Space Shuttle.
   c. GRISAFFE, S. J.: Review of Some Current NASA-Lewis Sponsored Coating Research.
   d. JONES, M. S.: Advances in Plasma Spray Coatings for High-Temperature.
   e. TUCKER, R. C.: Plasma-Deposited FeCrAlY Coatings.

Section II

Plasma Testing Literature

1. ADAMS, E. W.: Analysis of Quartz and Teflon Shields for a Particular Reentry Mission. Proceedings of 1961 Heat Transfer and Fluid Mech. Inst. Pasadena, California, June, 1961.

2. AHOUSE, D. R., and P. J. HARBOUR: The Flow Upstream of a Blunt Body in Hypersonic Flow from Continuum to Nearly Free Molecule Conditions. Book of Abstracts, **1**, Sixth International Symposium on Rarefied Gas Dynamics, 1968.

3. ALLEN, H. J., and A. J. EGGERS, JR.: A Study of the Motion and Aerodynamic Heating of Ballistic Missiles Entering the Earth's Atmosphere at High Supersonic Speeds. NACA Report 1381, National Committee for Aeronautics, 1958.

4. ALPHER, R. A., and D. R. WHITE: Optical Refractivity of High-Temperature Gases, **I**, Effects Resulting from Dissociation of Diatomic Gases, Phys. Fluids, March-April, 1959.

5. ANDERSON, L. A., and R. E. SHELDAHL: Flow-Swallowing Enthalpy Probes in Low-Density Plasma Streams. AIAA Paper No. 68-390, April, 1968.

6. ANONYMOUS: Air Arc Test Facilities of the Space Sciences Laboratory. General Electric SSL Report 4, April, 1961.

7. ANONYMOUS: Plastic Thermocouple. Tech. Bull. 161, Nanmac Corporation, Indian Head, Maryland (No Date).

8. ANONYMOUS: Wave Superheater Hypersonic Tunnel. Cornell Aeronautical Laboratory Report, December, 1960.

9. ANTHONY, F., and A. MISTRETTA: Investigation of Feasibility of Utilizing Available Heat Resistant Materials for Hypersonic Leading Edge Applications. ASD-TR-59-744, U.S. Air Force, 1961.

10. ARD, W. B.: Electron Temperature Measurements Above $100,000°$ K Using Radiation at the Electron Cyclotron Frequency. Paper No. 67, Temperature, Its Measurement and Control in Science and Industry, **3**, Part 1, Fourth Symposium, New York: Reinhold Publishing Company, 1962.

11. ASHKENAS, H.: On Rotational Temperature Measurements in Electron-Beam Excited Nitrogen. Phys. Fluids, **10**, 2509 (1967).

12. AVERSEN, JOHN C.: Effectiveness of Solar Radiation Shields for Thermal Control of Space Vehicles Subjected to Large Changes in Solar Energy. NASA Washington Symposium on Thermal Radiation of Solids, and National Aeronautics and Space Administration, 1965.

13. BARKAN, P., and A. WHITMAN: An Uncooled, Rapid Response Probe for Measuring Stagnation Pressures in High Velocity Arc Plasma. ARL-64-192, U.S. Air Force, 1962.

14. BARR, R., R. R. BOEDEKER, R. M. DELANEY, P. D. DOHERTY, D. J. MANSON, and A. H. WEBER: Temperature Measurements in High-Temperature Gases Using Thermal Neutron Scattering. Bull. Amer. Phys. Soc., **6** (1961).

15. BARRIAULT, R. S., and J. YOS: Analysis of the Ablation of Plastic Heat Shields that Form a Charred Surface Layer. J. Amer. Rocket Soc., **30**, No. 9 (1960).

16. BATCHELOR, J.: Behavior of Nozzle Materials Under Extreme Rocket Motor Environment. QPR No. 2, Atlantic Research Corp., December, 1965.

17. BATCHELOR, J., N. VASILEFF, S. McCORMICK, and E. OLCOTT: Development and Evaluation of Solid Propellant Rocket Motor Case Insulating Materials Systems. WADD TR-60-109 Part I, U.S. Air Force, 1960.

18. BATCHELOR, J., N. VASSILEFF, S. McCORMICK, and E. OLCOTT: Insulation Materials for Solid Propellant Rocket Motors. WADD TR-60-109, Part 2, U.S. Air Force, 1961.

19. BECKWITH, E., and N. B. COHEN: Application of Similar Solutions to Calculation of Laminar Heat Transfer on Bodies with Yaw and Large Pressure Gradient in High Speed Flow. NASA TN D-625, National Aeronautics and Space Administration, 1961.

20. BETHE, H. A., and MAC C. ADAMS: A Theory for the Ablation of Glassy Materials. J. Aerospace Sci., **26**, June (1959).

21. BIEKOWSKI, G. K., and P. J. HARBOUR: Structure of Electron-Beam Generated Plasma. Presented at Rarefied Gas Dynamics, Sixth Symposium, 1968.

22. BLACKSHEAR, P. J., and F. D. DORMAN: Heat-Sensing Probe and Process. U.S. Patent No. 3, 296, 865, (Diluent Probe), January 10, 1967.

23. BLACKSHEAR, P. J.: Sonic Flow Orifice Temperature Probe for High Gas Temperature Measurements. NACA TN 2167, National Advisory Committee for Aeronautics, 1950.

24. BLYTHE, P. A.: Nonequilibrium Flow Through a Nozzle. Fluid Mech. Papers, 17 (1963).

25. BOATRIGHT, W. B., D. I. SEBACHER, and R. W. GUY: Review of Testing Techniques and Flow Calibration Results for Hypersonic Arc Tunnels. AIAA Paper No. 68-379, April, 1968.

26. BOATRIGHT, W. B., R. B. STEWART, and D. I. SEBACHER: Testing Experience and Calibration Experiments in a Mach Number 12, 1-Foot Hypersonic Arc Tunnel. Presented at Third Hypervelocity Techniques Symposium, Denver, Colo., March, 1964.

27. BOGDAN, L., and K. C. HENDERSHOT: Density and Temperature Measurements in the Base Region of a Clustered Rocket Model Using an Electron Beam. Presented at Second International Congress on Instrumentation in Aerospace Simulation Facilities, August, 1966.

28. BOHM, D.: Minimum Ionic Kinetic Energy for a Stable Sheath. The Characteristics of Electrical Discharges in Magnetic Fields, New York: McGraw-Hill Book Co., 1949.

29. BOND, A., B. RASHINS, and R. LEVIN: Experimental Ablation Cooling. NACA Res. Memo. L 58 E 15a, National Advisory Committee for Aeronautics, 15 July, 1958.

30. BONIN, J., and C. PRICE: Thermal Protection of Structural, Propulsion, Temperature Sensitive Materials for Hypersonic and Space Flight. WADC TR-59-366, Part 1, U.S. Air Force, 1959.

31. BONIN, J., C. PRICE, and D. TAYLOR: Determination of Factors Governing Selection and Application of Materials for Ablation Cooling of Hypervelocity Vehicles. WADC TR 59-87, Part 1, U.S. Air Force, 1959.

32. BOWERS, D. A.: Leading Edges Development-Dyna Soar. D 2 80085, The Boeing Co., June, 1963.

33. BOYER, A. G., and E. P. MUNTZ: Experimental Studies of Turbulence Characteristics in the Hypersonic Wake of a Sharp Slender Cone. The Fluid Physics of Hypersonic Wakes, AGARD CP No. 14, May, 1967.

34. BRATT, L., and D. CHAMBERLAIN: High-Temperature Synthesis of New, Thermally Stable Chemical Compounds. ASD TR-59-345, U.S. Air Force, 1959.

35. BROCKLEHURST, R. E.: High Intensity Gamma Ray Dosimetry. ASTM Special Technical Publication, American Society for Testing and Materials, 1960.

36. BRODING, W. C., F. W. DIEDERICH, and P. S. PARKER: Structural Optimization and Design Based on a Reliability Design Criterion. Presented at AIAA Shell Conference, Palm Springs, California, April, 1963.

37. BRODING, W. C., T. R. MUNSON, and R. H. WATSON: Aerothermal Structural Analysis of Carbonaceous Heat Shield Materials. RAD-RM-64-38, July, 1964.

38. BROGAN, T.: The Electric Arc Wind Tunnel A Tool for Atmospheric Re-entry Research. J. Amer. Rocket Soc., 29, 648 (1958).

39. BROGAN, T. R.: Electric Arc Gas Heaters for Re-entry Simulation and Space Propulsion. AVCO-Everett Research, Rept. No. 35, 1958.

40. BROWN, J. D., and F. A. SHUKIS: An Approximate Method for Design of Thermal Protection Systems. AVCO RAD TM-62-3. Presented at IAS 30th Annual Meeting in New York, January 21, 1962.

41. BROWN, S. C., G. BEKEFI, and R. E. WHITNEY: Infrared Interferometer for the Measurement of High Electron Densities. J. Opt. Soc. Amer., 53 (1963).

42. BRUNDIN, C. L., and L. TALBOT: The Application of Langmuir Probes to Flowing Ionized Gases. AGARD Report 478, September, 1964.

43. BUDNICK, L., H. HALLE, and C. PRICE: Performance Characteristics of High-Temperature Materials Exposed to Reentry Conditions Simulated by an Air-Stabilized Arc. WADD TR 61-81, Part 1, U.S. Air Force, 1961.

44. BUHLER, R., and D. CHRISTENSEN: Arc Jet Measurements Related to Ablation Test Validity. ASTM Special Tech. Publ. 279, p. 107, Reinforced Plastics for Rockets and Aircraft, American Society for Testing and Materials, 1959.

45. BUHLER, R., D. CHRISTENSEN, and S. GRINDLE: Effects of Hyperthermal Conditions on Plastics Ablation Materials. ASD-TR-61-304, U.S. Air Force, 1961.

46. BUNDY, and STRONG: Physical Measurements in Gas Dynamics and Combustion. Princeton Series in High-Speed Aerodynamics, 1956.

47. BUTEFISCH, K.: Investigation of Hypersonic Non-Equilibrium Rarefied Gas Flow Around a Circular Cylinder by the Electron Beam Technique. Presented at Rarefied Gas Dynamics Sixth Symposium, 1968.

48. BUTLER, C.: Image Furnace Research. International Symposium on High-Temperature Technology, New York, McGraw-Hill Publishing Company, 1959.

49. BUTZ, J.: Growth Potential Defined for Heat Sink, Ablation Shields. Aviation Week Space Tech., **71**, 69 (1959).

50. BYRON, S., and R. M. SPONGBERG: Gasdynamic Instrumentation of High Enthalpy Flows. IEEE Transactions on Nuclear Science, NS-11, January, 1964.

51. CAMAC, M.: Flow Field Measurements with an Electron Beam. AIAA Paper 68-7222, 1968.

52. CANN, C. L., J. M. TEEM, R. BUHLER, and L. K. BRANSON: Magnetogasdynamic Accelerator Techniques. AEDC-TDR-62-145, U.S. Government, July, 1962.

53. CARDEN, W. H.: Heat Transfer in Non-Equilibrium-Dissociated Hypersonic Flow with Surface Catalysis and Second-Order Effects. AIAA Journal, **4**, October, 1966.

54. CARNEVALE, E. H., H. L. POSS, and J. M. YOS: Ultrasonic Temperature Determinations in a Plasma. Paper 89, Temperature, Its Measurement and Control in Science and Industry, **2**, Part 2, New York: Reinhold Publishing, 1962.

55. CARSLAW, H. S., and J. C. JAEGER: Conduction of Heat in Solids. Oxford University Press, 1959.

56. CARSWELL, A. I., M. P. BACHYNSKI, and G. G. CLOUSZIER: Microwave Measurements of Electromagnetic Properties of Plasma-Flow Fields. AIAA Paper 63-385, 1963.

57. CASON, C. M., and T. A. BARR: The Argma Plasma-Jet Facility. AD-263-470, U.S. Army, 1960.

58. CENTER, R. E.: Measurement of Shock Wave Structure in Helium-Argon Mixtures. Phys. Fluids, **10**, 1777 (1967).

59. CHEN, C. J.: Velocity Surveys in Arc-Jet Tunnel Using Micron Size Tracer Particles. Santa Ana, Calif., Plasmadyne Corporation (No Date).

60. CHENG, D. Y., and P. L. BLACKSHEAR: Factors Influencing the Performance of a Fast-Response, Transpiration-Cooled, High-Temperature Probe. AIAA Paper No. 65-359, July, 1965.

61. CHRISTENSEN, D., and R. BUHLER: Arc-Jet Tunnel Development and Calibration for Parabolic Re-Entry Simulation. Final Summary Report IFR O 11-1872 on Contract DA-04-495-506-ORD-1872, Santa Ana, Calif., Plasmadyne Corp., 1961.

62. COBINE, J. D., and E. E. BURGER: Ionization Gauge for Transient Gas Pressures. Rev. Sci. Instr., **32**, No. 6 (1961).

63. COFFMAN, J., G. KIBLER, T. RIETHOF, and A. WATTS: Carbonization of Plastics and Refractory Materials Research. WADD TR-60-646, Part 1, U.S. Air Force, 1961.

64. COHEN, I. M.: Asymptotic Theory of Spherical Electrostatic Probes in a Slightly-Ionized, Collision-Dominated Gas. Phys. Fluids, **6**, 1492—1499 1963.

65. COLBURN, A. P.: A Method of Correlating Forced Convection Heat Transfer and a Comparison with Fluid Friction. Transactions of the AIChE, **29** (1933).,

66. COOKSON, T. S., P. G. DUNHAM, and J. K. KILHAM: Stagnation Point Heat Flow Meter Instr., **42**, April (1965).
67. COOPER, L. Y.: On the Magnetically Driven T-Type Shock Tube of Rectangular Geometry. AFCRL-687, U.S. Air Force, 1961.
68. COOTE, M.: A Preliminary Design Method for Estimating the Drag of Cones. AVCO RAD Report L 343-MAC-58, 1958.
69. CORDERO, J., F. W. DIEDERICH, and H. HURWICZ: Aerothermodynamic Test Techniques for Re-Entry Structures and Materials. Aerospace Eng. **22**, January (1963).
70. CORRAIN, S.: Extended Applications of the Hot Wire Anemometer. NACA TN 1864, National Advisory Committee for Aeronautics, 1949.
71. COUDEVILLE, H., I. I. VIVIAND, M. RAFFIN, and E. A. BRUN: An Experimental Study of Wakes of Cylinders at Mach 20 in Rarefied Gas Flows. Book of Abstracts, **I**, Sixth International Symposium on Rarefied Gas Dynamics, 1968.
72. CRABOU, R., C. PANNABECKER, A. PLATE, and R. RICE: The Determination of Hypersonic Drag Coefficients for Cones and Spheres. RAD TM-65-32, 1965.
73. CRITES, R. C., and P. CZYSZ: Inlet and Test Section Diagnostics Using a Miniature Mass Flow Probe in Hypersonic Impulse Tunnel. AIAA Paper No. 68-389, April, 1968.
74. CUNNINGHAM, J. W., C. H. FISHER, and L. L. PRICE: Density and Temperature in Wind Tunnels Using Electron Beams. IEEE Transactions on Aerospace and Electronic Systems, **2**, 269 (1967).
75. DAVIS, R. M., and C. MILEWSKI: High-Temperature Composite Structure. ASD TDR-62-418, U.S. Air Force, 1962.
76. DE LEEUW, J. H.: Electrostatic Plasma Probes. AIAA Paper No. 63-370, Fifth Biennial Gas Dynamics Symposium, August 14—16, 1963.
77. DEMETRIADES, A., and E. L. DOUGHMAN: Langmuir Probe Diagnosis of Turbulent Plasmas. AIAA Journal, **4** (1966).
78. DEMETRIADES, A., and E. L. DOUGHMAN: Langmuir-Probe-Measurement Method for Turbulent Plasmas. Phys. Fluids, **8** (1965).
79. DIACONIS, N., J. FANUCCI, and G. SUTTON: The Heat Protection Potential of Several Ablation Materials for Satellite and Ballistic Re-Entry into the Earth's Atmosphere. Planetary Space Sci., **4**, 463 (1963).
80. DICKERMAN, P. J.: Optical Spectrometric Measurements of High-Temperatures. University of Chicago, 1961.
81. DIEDERICH, F. W., W. C. BRODING, A. S. HANAWALT, and R. SIRUALL: Reliability as a Thermostructural Design Criterion. Transactions of 7th Symposium on Ballistic Missile and Space Technology, Air Force Academy, August, 1962.
82. DIEKE, G. H.: Spectroscopy of Combustion. In Physical Measurements in Gas Dynamics and Combustion, Princeton Series in High Speed Aerodynamics, **IX** (1959).
83. DIMMOCK, T. H.: Influence of Ions on Rocket Combustion. AFOSR-63 Final Report on Project (42205) 204, U.S. Air Force, 1963.
84. DORMER, G., B. PAYNE, and E. PIKE: Manufacturing Methods for High-Temperature Coating Of Large Molybdenum Parts. AFML TR-65-28, U.S. Air Force, 1965.
85. DOUGAL, A., and R. GRIBBLE: Research on Plasma Diagnostic Methods for High-Temperature Plasma Research. ARL 65-270, U.S. Air Force, 1965.
86. DOYLE, C.: Evaluation of Experimental Polymers. WADC TR-59-136, U.S. Air Force, 1959.
87. EAST, L. F.: Measurement of Skin Friction at Low Subsonic Speeds by the Razor Blade Technique. ARC-R & M-3525, Aeronautical Research Council, 1968.
88. ECKERT, E. R. G.: Engineering Relations for Heat Transfer and Friction in High Velocity Laminar and Turbulent Boundary Layer Flow over Surfaces with Constant Pressure and Temperature. ASME Paper No. 55-A-31, 1955.

89. EDMONSON, R. B., W. R. THOMPSON, and A. L. HINES: Thermodynamic Temperature Probe. Am. Rocket Soc. Preprint No. 1431-60, December, 1960.

90. EPSTEIN, G., and H. KING: Plastics for Rocket Motor Nozzles. Ind. Eng. Chem., **9** (1960).

91. EPSTEIN, G., and J. WILSON: Reinforced Plastics for Rocket Motor Applications. S. P. E. Journal, **6** (1959).

92. ESKER, D. W.: A Probe for Total-Enthalpy Measurements in Arc-Jet Exhausts. AIAA Journal, **5**, August (1967).

93. FAY, J. A., and N. H. KEMP: Theory of Stagnation-Point Heat Transfer in a Partially Ionized Diatomic Gas. Presented at IAS Annual Meeting, New York, January 21—23, 1963.

94. FAY, J. A., and F. R. RIDDELL: Theory of Stagnation-Point Heat Transfer in Dissociated Air. J. Aerospace Sci., **25**, February (1968).

95. FINGERSON, L. M.: Research on The Development and Evaluation of a Two-Sensor Enthalpy Probe. Report No. ARL 64-161, U.S. Air Force, 1964.

96. FINGERSON, L. M., and P. L. BLACKSHEAR: Characteristics of Heat-Flux Probe for Use in High-Temperatures Atmospheres. International Developments in Heat Transfer, Part IV, 1961.

97. FINGERSON, L. M., and P. L. BLACKSHEAR: Some New Measuring Techniques in High-Temperature Gases. Combustion Lab., Univ. of Minnesota, Tech. Rep. 61-3, 1961.

98. FIOK, E. F. and A. I. DAHL: The Measurement of Gas Temperatures by Immersion-Type Instruments. J. Amer. Rocket Soc., **23**, No. 3 (1953).

99. FISCHER, H., and L. C. MANSUR: Conference on Extremely High-Temperatures. New York: John Wiley and Sons, 1958.

100. FLEDDERMAN, R. G., and H. HURWICZ: Analysis of Transient Ablation and Heat Conduction Phenomena at a Vaporizing Surface. Chemical Engineering Symposium Series, **57**, 24 (1960).

101. FLOCK, J. L., and R. R. HECK: Operational Experiences and Preliminary Results of Total Enthalpy Probe Measurements in the AFFDL 50-Megawatt Electrogasdynamics Facility. USAF Report No. FDM-TM-68-2, U.S. Air Force, 1968.

102. FONTIJN, A., D. E. ROSNER, and S. C. KURZIUS: Chemical Scavenger Probe Studies of Atom and Excited Molecule Reactivity in Active Nitrogen from a Supersonic Stream. Can. J. Chem., **42** (1964).

103. FORBES, S. G., S. C. SLATTERY, and R. F. KEMP: Beam Diagnostic Studies. Presented at USAF-NASA Joint Meeting on Electrostatic Propulsion, April 24—26, 1961.

104. FREEMAN, M. P.: Plasma Jet Diagnosis Utilizing the Ablating Probe. Temperature — Its Measurement and Control in Science and Industry, **3**, New York: Reinhold Publishing Corp., 1962.

105. FRENCH, J. B.: Langmuir Probes in a Flowing Low-Density Plasma. Institute of Aerophysics, UTIA Report No. 79, Univ. of Toronto, August, 1961.

106. FRUCHTMAN, I.: Temperature Measurement of Hot Gas Stream. AIAA Journal, **I**, August, 1963.

107. GADAMER, E. O.: Measurement of the Density Distribution in a Rarefied Gas Flow Using the Fluorescence Induced by a Thin Electron Beam. UTIA Report 83, AFOSR TN 60-500, U.S. Government, 1962.

108. GARSTANG, R. H.: Forbidden Transitions in Atomic and Molecular Processes. Edited by D. R. Bates, Academic Press, 1962.

109. GAYDON, A. G., and H. G. WOLFHARD: Flames, Their Structure, Radiation, and Temperature. Chapman and Hall Ltd., 1953.

110. GAYELEY, C.: Theoretical Evaluations of the Turbulent Skin Friction Coefficient and Heat Transfer on a Cone in Supersonic Flight. Report No. R 49 A 052, U.S. Army, 1949.

111. GEORGIEV, S., H. HIDALGO, and M. ADAMS: On Ablation for the Recovery of Satellites. Heat Transfer and Fluid Mechanics Institute Papers, p. 171, Stanford University Press, 1959.

112. GLASER, P.: Imaging Furnace Developments for High-Temperature Research. Presented at Electrochemical Society Meeting, Philadelphia, Pa., May, 1959.

113. GLASS, I. I., and J. G. HALL: Handbook of Supersonic Aerodynamics. 18, NAVORD Rept. 1488, 6, U.S. Government Printing Office, Section 7, 1959.

114. GORDON, G. M., and D. A. BROWN: Tungsten and Rocket Motors. Stanford Research Inst. (No Date.)

115. GRABOW, R.: Mass Loss Calculations for Trajectory Analysis. AVCO/RAD-TR-63-154, 1963.

116. GRABOW, R., and P. LYNCH: Preliminary Design Curves of Aerodynamic Heating and Skin Friction for Conical Reentry Vehicles. AVCO/RAD-TR-64-79, 1964.

117. GREENSHIELDS, D. H.: Spectrographic Temperature Measurements in Carbon-Arc-Powered Air Jet. NASA TN D-169, National Aeronautical and Space Administration, 1959.

118. GREY, J.: Cooled Probe Diagnostics of Dense Plasma Mixing and Heat Transfer Processes. AIChE, Preprint 9C, November, 1967.

119. GREY, J.: Enthalpy Probes for Arc Plasmas-First Status Review. Report to Section 3, Subcommittee VI, Committee E-21, ASTM, April 12, 1966.

120. GREY, J.: Enthalpy Probes for Arc Plasmas-Second Status Review. Report to Section 3, Subcommittee VI, Committee E-21, ASTM, Toronto, Canada, May 3, 1967.

121. GREY, J.: Sensitivity Analysis for the Calorimetric Probe. Rev. Sci. Instr., 34, August (1963).

122. GREY, J.: Thermodynamic Methods of High-Temperature Measurement. ISA Transactions, 4, April (1965).

123. GREY, J., and P. F. JACOBS: A Calorimetric Probe for the Measurement of High Gas Temperatures. AEL Rept. No. 602, Princeton Univ., April, 1962.

124. GREY, J., and P. F. JACOBS: Cooled Electrostatic Probe. AIAA Journal, 5, January (1967).

125. GREY, J., and P. F. JACOBS: Experiments on Turbulent Mixing in a Partially Ionized Gas. AIAA Journal, 2, March (1964).

126. GREY, J., and F. SHERMAN: Calorimetric Probe for the Measurement of Extremely High-Temperatures. Rev. Sci. Instr., 33, July (1962).

127. GREY, J., M. P. SHERMAN, and P. F. JACOBS: A Collimated Total Radiation Probe for Arc Jet Measurements. IEEE Transactions on Nuclear Science, NS-11, January (1964).

128. GREY, J., M. P. SHERMAN, P. M. WILLIAMS, and D. B. FRADKIN: Laminar Arc Jet Mixing and Heat Transfer: Theory and Experiments. AIAA Journal, 4, June (1966).

129. GRIEM, H. R.: Plasma Spectroscopy. Paper for the Fifth International Conference on Ionization Phenomena in Gases, II, Munich, 1961.

130. GROSS, J. J., D. J. MASSON, and C. GAZLEY, JR.: General Characteristics of Binary Boundary Layer with Applications to Sublimation Cooling. Rand Rep. P-1371, May 8, 1958.

131. GROSSE, A. V., and C. S. STROKES: Plasma Jet Chemistry. Temple Univ., June, 1963.

132. GRUN, A. E.: A Gasdynamic and Spectroscopic Study of the Excitation of Gas Streams. Z. Naturforschg., 9a, 833 (1954).

133. GRUN, A. E., E. SCHOPPER, and B. SCHUMACHER: Application of Intense Corpuscular Beams for the Excitation of Gases. Z. Angew Phys., 6, 198 (1954).

134. GRUN, A. E., E. SCHOPPER, and B. SCHUMACHER: Electron Shadowgraphs, and After glow Picture of Gas Jets at Low Densities. J. Appl. Phys., 24, 1527 (1953).

135. HALBACH, C. R., and L. FREEMAN: The Enthalpy Sensor — A High Gas Temperature Measuring Probe. Report MR 20, 331, The Marquardt Corporation, June, 1965.

136. HALL, J. G., and C. E. TREANOR: Nonequilibrium Effects in Supersonic-Nozzle Flows. AGARD, December, 1967.

137. HALLE, H., and C. PRICE: Development of Test Facilities for Studies in Hypersonic Range. WADD TR 60-414, Part 1, U.S. Air Force, 1960.

138. HANST, P., and A. WALKER: The Infrared Emission Spectra of Plastics Ablating in a Low Enthalpy Air Stream: Measurements of Surface Temperatures and Temperature Profiles Behind the Surfaces. WADD TR 60-101, U.S. Air Force, 1960.

139. HARBOUR, P. J.: Absolute Determination of Flow Parameters in a Low Density Hypersonic Tunnel. Presented at Rarefied Gas Dynamics Sixth Symposium, 1968.

140. HARBOUR, P. J., and J. H. LEWIS: Preliminary Measurements of the Hypersonic Rarefied Flow Field on a Sharp Flat Plate Using an Electron Beam Probe. Rarefied Gas Dynamics, **II**, 1031, 1967.

141. HARTUNIAN, R. A.: Theory of a Probe for Measuring Local Atom Concentrations in Hypersonic Dissociated Flows at Low Densities. DCAS-TDR-62-101, Aero. Space Report No. TDR-930 (2230-06) TN-2, U.S. Government, 1962.

142. HASS, F. C.: An Evaporating Film Calorimetric Enthalpy Probe. Report No. AD-1651-Y-1, Cornell Aeronautical Laboratory, February, 1963.

143. HASS, F. C., and F. A. VASSALLO: Measurement of Stagnation Enthalpy in a High Energy Gas Stream. Chemical Engineering Progress Symposium Series 41, **59**, AIChE (1963).

144. HAVERSTRAW, R.: High-Temperature Extrusion Lubricants. ER 5822, U.S. Air Force, April, 1964.

145. HECHT, G.: A Novel, Near Infrared, Two-Wavelength Pyrometer. Presented at Symposium on Temperature, Its Measurement and Control in Science and Industry, Columbus, Ohio, March 27—31, 1961.

146. HICKMAN, R. S.: An Experimental Study of Hypersonic Rarefied Flow Over a 10° Conc. Rarefied Gas Dynamics, **II**, 1967.

147. HICKMAN, R. S.: Electron Beam Flow Field Visualization. Presented to A.P.S. Fluid Dynamics Div., Cleveland, Ohio, November, 1965.

148. HICKMAN, R. S.: Hypersonic Transitional Flow at the Leading Edge of a Sharp Flat Plate. Presented at Rarefied Gas Dynamics Sixth Symposium, 1968.

149. HICKMAN, R. S.: Rotational Temperature Measurements in Nitrogen Using an Electron Beam. USCAE 104, Univ. of Southern California, September, 1966.

150. HIDALGO, H.: Ablation of Glassy Materials Around Blunt Bodies of Revolution. J. Am. Rocket Soc., **30**, 806, September (1960).

151. HIESTER, N. K., and CARROLL F. CLARK: Feasibility of Standard Evaluation Precedures for Ablating Materials. NASA CR-379, National Aeronautics and Space Administration, 1965.

152. HILL, W. E.: Plasma Temperature, Relative Intensity of Spectral Lines. Paper 58 in Temperature, Its Measurement and Control in Science and Industry, **3**, Part 1 (1962).

153. HOERNER, S.: Fluid Dynamic Drag. Published by Author, Midland Park, New Jersey, 1958.

154. HOOTON, N.: Materials Property Data. QPR No. 1, Bendix Corp., July, 1961.

155. HOTTEL, H. C., G. C. WILLIAMS, and W. P. JENSEN: Optical Methods of Measuring Plasma Jet Temperatures. WADD TR-60-676, U.S. Air Force, 1961.

156. HOURT, W.: Plastics as Heat Insulators in Rocket Motors. Ind. Eng. Chem., **9** (1960).

157. HUBER, F. J. A.: Probes for Measuring Mass Flux, Stagnation Point Heating and Total Enthalpy of High-Temperature Hypersonic Flows. AIAA Preprint No. 66-750, September, 1966.

158. HUNTER, W. W., JR.: Rotational Temperature Measurements from 300° K to 1000° K with Electron Beam Probe. I.S.A. Preprint 16.12-4-66, October, 1966.

159. HURLBUT, F. C.: An Electron Beam Density Probe for Measurements in Rarefied Gas Flows. WADC-TR-57-644, U.S. Air Force, 1957.

160. HURTT, W. W., G. CUNNINGTON, A. FUNAI, J. GRAMMER, and R. KARLAK: Program ASTEC (Advanced Solar Turbo Electric Concept). AFAPL TR-65-53, U.S. Air Force, 1966.

161. INCROPERA, F. P.: Temperature Measurement and Internal Flow Heat Transfer Analysis for Partially Ionized Argon. Technical Report No. SU 247-11, Department of Mechanical Engineering, Stanford University, August, 1966.

162. INCROPERA, F. P., and G. LEPPERT: Investigation of Arc Jet Temperature-Measurement Techniques. ISA Transactions, **6**, January (1967).

163. JACOBS, P. F., and J. GREY: Electron-Heavy Particle Nonequilibrium in a Dense Argon Plasma. Report No. ARL 66-0143, U.S. Air Force, 1966.

164. JACOBS, P. F., and J. GREY: A Criterion for Electron-Heavy-Particle Nonequilibrium in a Partly-Ionized Gas. AIAA Preprint 66-192, March, 1966.

165. JAHN, R. E.: Spectroscopic Measurements of the Temperature of Plasma Jets. Presented at Fifth International Conference on Ionization Phenomena in Gases, 1962.

166. JAMES, W. R.: In Vacuo Trajectory Influence Coefficients and Deployment Studies. Report No.: E 210-64-WRJ 205, AVCO/RAD, 1965.

167. JOHN, R., and W. L. BADE: Recent Advances in Electric Arc Plasma Generation Technology. J. Amer. Rocket Soc. **31**, No. 1 (1961).

168. JOHN, R., and W. BADE: Testing of Reinforced Plastics Under Simulated Re-Entry Conditions. ASTM Special Tech. Publ. 279, P. 93, Symposium on Reinforced Plastics for Rockets and Aircraft American Society for Testing and Materials, 1959.

169. JOHN, R., and J. RECESSO: Ablation Characteristics of a Subliming Material Using Arc Heated Air. J. Amer. Rocket Soc., **9** (1959).

170. JOHNSON, D. H.: Nonequilibrium Electron Temperature Measurement in a Supersonic Arc Jet Using a Cooled Langmuir Probe. Presented at 26th Supersonic Tunnel Association Meeting, Ames Research Center, May 16—18, 1967.

171. JOHNSON, E. O., and L. MALTER: A Floating Double Probe Method for Measurements in Gas Discharges. Phys. Rev., **80**, No. 1 (1950).

172. JULIUS, J. D.: Measurements of Pressure and Local Heat Transfer on a 20-degree Cone at Large Angles of Attack up to 20-degrees for a Mach Number of 4.95. NASA TN D-179, National Aeronautics and Space Administration, 1959.

173. KANZLARICH, J. J.: Ablation of Reinforced Plastic for Heat Protection. J. of Appl. Mechanics, **32**, No. 1 (1965).

174. KILBURG, R. F.: A High Response Probe for Measurement of Total Temperature and Total Pressure Profiles Through a Turbulent Boundary Layer with High Heat Transfer in Supersonic Flow. AIAA Paper No. 68—374, April, 1968.

175. KIMURA, I., and A. KANZAWA: Experiments on Heat Transfer to Wires in a Partially Ionized Argon Plasma. AIAA Journal, **3** (1965).

176. KLEIN, A. F.: A Survey of Optical Interferometry as Applied to Plasma Diagnostics. AIAA Paper 63-377. Presented at Fifth Biennial Gas Dynamics Symposium, August 14—16, 1963.

177. KOPSEL, M., and M. RICHTER: Research on Method for Measuring Temperatures Between 8000 Degrees and 30,000 Degrees Kelvin. ASD-TDR-62-916, U.S. Air Force, 1962.

178. KOSTKOWSKY, H. J.: The Accuracy and Precision of Measuring Temperatures Above 1000° K. Presented at International Symposium on High-Temperature Technology, Asilomar Conference Ground, October 6—9, 1959.

179. KOUBEK, F., and A. TIMMINS: High-Temperature Facilities at the Naval Ordnance Laboratory. NAVWEPS Report 7315, U.S. Navy, 1960.

180. Kovasznay, L. S. G.: The Hot Wire Anemometer. Acta, Tech. Acad. Sci Hung., **50**, 131 (1965).

181. Kovasznay, L. S. G.: Turbulence Measurement. High Speed Aerodynamics and Jet Propulsion, Princeton Univ. Press, 1954.

182. Krause, L. N., D. R. Buchele, and I. Warshawsky: Measurement Techniques for Hypersonic Propulsion. NASA TM X-52299, National Aeronautics and Space Administration, 1967.

183. Krause, L. N., G. E. Glawe, and R. C. Johnson: Heat Transfer Devices for Determining the Temperature of Flowing Gases. Temperature — Its Measurement and Control in Science and Industry, New York: Reinhold Publishing Co., 1962.

184. Krouse, J.: Characteristics in a High-Enthalpy, Low-Density Wind Tunnel. DTMB-1921, DTMB-AERO-1076, David Taylor Model Basin, September, 1964.

185. Kubanek, G. R., and W. H. Gauvin: Plasma Jet Research Facility for Solid-Gas Heat Transfer Studies. Technical Report No. 466, Pulp and Paper Research Institute of Canada, 1966.

186. Ladenbery, R., et al.: Physical Measurements in Gas Dynamics and Combustion. Princeton, N. J., Princeton Univ., 1964.

187. Lally, F. T., and D. Laverty: Carbides for Solid Propellant Nozzle Systems. AFRPL TR-68-164, U.S. Air Force, 1968.

188. Lam, S. H.: A General Theory for the Flow of Weakly Ionized Gases. AIAA Journal, **2**, February (1964).

189. Landau, H. G.: Heat Conduction in a Melting Solid. Quart. J. Appl. Math. and Phys., **8**, April (1950).

190. Landry, B. E.: Nose Cap Developments Tests. D 2 80083, The Boeing Co., September, 1963.

191. Langmuir, I.: The Pressure Effect and Other Phenomena in Gaseous Discharges. J. Franklin Inst., **196** (1923).

192. Laszlo, T.: Measurement of Emission of Radiant Energy From Samples Heated by the Plasma Jet. AVCO RAD Technical Memorandum 60-14, 31 March, 1960.

193. Laufer, J.: The Structure of Turbulence in Fully Developed Pipe Flow. NACA TN 2954, National Advisory Committee for Aeronautics, 1953.

194. Leadon, B. M., and C. J. Scott: Mass Transfer Cooling at Mach Number 4.8. J. Aeron. Sci., **25**, January (1958).

195. Leadon, B. M., and C. J. Scott: Measurement of Recovery Factor and Heat Transfer Coefficients with Transpiration Cooling in a Turbulent Boundary Layer at $M = 3.0$ Using Air and Helium as Coolants. Research Rep. 126, Institute of Technology University of Minnesota, February, 1956.

196. Lees, L.: Convective Heat Transfer with Mass Addition and Chemical Reactions. Combustion and Propulsion, Third AGARD Colloquim, New York: Pergamon Press, 1959.

197. Lees, L.: Laminar Heat Transfer Over Blunt-Nosed Bodies at Hypersonic Flight Speeds. Jet Propulsion, **26**, No. 4 (1956).

198. Lemay, A.: The CARDE Turbulent Hypersonic Point Measurement Program. CARDE TN 1772/67, October, 1967.

199. Leonard, D. A., and J. C. Keck: Schileren Photographs of Projectile Wakes Using Resonance Radiation. J. Amer. Rocket Soc., **32**, No. 7 (1962).

200. Li, Y. T.: Dynamic Pressure Measuring Systems for Jet Propulsion Research. J. Amer. Rocket Soc., **23**, No. 3 (1953).

201. Ling, S. C.: Heat Transfer Characteristics of Hot-Film Sensing Elements Used in Flow Measurements. J. of Basic Eng. 82, 1966.

202. Linnell, R. D.: Incompressible Newtonian Hypersonic Flow Around a Sphere. Research Note 5, Convair Scientific Research Laboratory, General Dynamics, June, 1957.

203. LINNELL and BAILEY: Similarity Rule Estimation Methods for Cones and Parabolic Noses. J. Aeron. Sci., **23**, No. 8 (1956).

204. LLINAS, J.: Electron Beam Measurements of Temperature and Density in the Base Region of a Clustered Rocket Model. AIAA Paper 68-236, March, 1968.

205. LOCHTE-HOLTGREVEN, W.: Ionization Measurements of High-Temperatures. Temperature, Its Measurement and Control in Science and Industry, **2**, New York: Reinhold Publishing Co., 1955.

206. LOCHTE-HOLTGREVEN, W.: Production and Measurements of High-Temperatures. Progress in Phys., **XXI** (1958).

207. LOCHTE-HOLTGREVEN, W., and H. MAECKER: Temperature Determination of Free Burning Arcs with the Aid of CN Bands. Zeit. für Phys., **105** (1937).

208. LORD, J.: Design and Performance Characteristics of an Automatic Brightness Pyrometer. Presented at Symposium on Temperature, Its Measurement and Control in Science and Industry, Columbus, Ohio, March, 1961.

209. LUCAS, W., and J. KINGSBURY: The ABMA Reinforced Plastics Ablation Program. Modern Plastics, **2** (1960).

210. MACARTHUR, R. C., L. M. STEVENSON, and J. BUDELL: Flow Visualization and Quantitative Gas Density Measurements in Rarefied Gas Flows. ASD-TDR-62-793, U.S. Air Force, 1962.

211. MADORSKY, S.: Thermal Degradation of Organic Polymers. SPE Journal, **7** (1961).

212. MADORSKY, S., and S. STRAUS: Thermal Degradation of Polymers at Temperatures up to 850° C. WADC TR 59-64, Part 1, U.S. Air Force, 1959.

213. MADORSKY, S., and S. STRAUS: Thermal Degradation of Polymers at Temperatures up to 1200° C. WADC TR 59-64, Part 2, U.S. Air Force, 1960.

214. MAGUIRE, L.: Density Effects on Rotational Temperature Measurements in Nitrogen Using the Electron Beam Excitation Technique. Presented at Rarefied Gas Dynamics Sixth Symposium, 1968.

215. MAGUIRE, L.: The Effective Spatial Resolution of the Electron Beam Fluorescence Probe in Helium. Rarefied Gas Dynamics, **2**, 1497, New York: Academic Press, 1967.

216. MAGUIRE, L., E. P. MUNTZ, and J. R. MALLIN: Visualization Technique for Low Density Flow Fields. IEEE Transactions on Aerospace and Electric Systems, **2**, 321 (1967).

217. MALLIARIS, A. C., et al.: Optical and Radar Observables of Ablative Materials. AFML TR-66-331, Part 1, U.S. Air Force, 1966.

218. MANOS, W., D. TAYLOR, and A. TUZZOLINO: Thermal Protection of Structural, Propulsion, and Temperature-Sensitive Materials for Hypersonic and Space Flight. WADC TR 59-366, Part 2, U.S. Air Force, 1960.

219. MARGRAVE, J. L.: Chemical Reactions in High-Temperature Systems. AFRPL TR-66-67, U.S. Air Force, 1966.

220. MARGRAVE, J. L.: Temperature Measurement in Physico-Chemical Measurements at High-Temperatures. London: Butterworth's Scientific Publications, 1959.

221. MARRONE, P. V.: Rotational Temperature and Density Measurements in Underexpanded Jets and Shock Waves Using an Electron Beam Probe. Phys. Fluids, **10**, 521 (1967).

222. MARSDEN, D. J.: The Measurement of Energy Transfer in Gas-Solid Surface Interactions Using Electron Beam Excited Emission of Light. Rarefied Gas Dynamics, **II**, 566 (1966).

223. MARTON, L., D. C. SCHUBERT, and S. R. MIELCZAREK: Electron Optical Studies of Low-Pressure Gases. NBS Monograph 66, National Bureau of Standards, August 16, 1963.

224. MASSEY, H. S. W., and E. H. S. BURHOP: Electronic and Ionic Impact Phenomena. Oxford University Press, 1951.

225. MASSIER, P. F., L. H. BACK, and E. J. ROSCHKE: Heat Transfer and Laminar Boundary Layer Distributions to an Internal Subsonic Gas Stream at Temperatures up to 13,900° R. Jet Propulsion Laboratory, 1968.

226. McCAA, D. J., and D. E. ROTHE: Fluorescence Spectra of Atmospheric Gases Excited by an Electron Beam. Presented at the Symposium on Molecular Structure and Spectroscopy, Ohio State Univ., September, 1968.

227. McCROSKEY, W. J.: Density and Velocity Measurements in High Speed Flows. AIAA Paper No. 68-392, April, 1968.

228. McGREGOR, W. K.: Spectroscopic Measurements in Plasmas. AIAA Paper No. 63-374. Presented at the Fifth Biennial Gas Dynamics Symposium, Northwestern Univ., August 14—16, 1963.

229. McNALLY, J. R., JR.: Role of Spectroscopy in Thermonuclear Research. J. Opt. Soc. Amer., 49 (1959).

230. MELNIK, W. L., and J. R. WESKE: Advances in Hot Wire Anemometry. Presented at the International Symposium on Hot Wire Anemometry, Univ. of Md., 20—21 March, 1967.

231. METCALF, S. C., D. C. LILLICRAP, and C. J. BERRY: Study of the Effect of Surface Temperature on the Shock-Layer Development Over Sharp-Edged Shapes in Low-Reynolds-Numbers High-Speed Flow. Presented at Rarefied Gas Dynamics Sixth Symposium, 1968.

232. MEYER, N.: Investigation of Greyrad Calorimetric Probe. BSE Thesis, College of Engineering, University of Cincinnati, 1966.

233. MILLER, N., and E. STRAUSS: Effects of Elevated Temperatures and Erosion on Reinforced Plastics Laminates. 13th Annual SPI Tech. Conf., p. 8-B, February, 1958.

234. MILLER, W.: Results of Heat Transfer Tests on Laminated Structures by the Use of an Oxy-Hydrogen Torch. NAVWEPS Report 7065, U.S. Navy, May, 1960.

235. MIXER, R., and C. MARYNOWSKI: A Study of the Mechanism of Ablation of Reinforced Plastics. WADC TR 59-668, Part 1, U.S. Air Force, 1960.

236. MOORE, D. W., JR.: A Pneumatic Method for Measuring High-Temperature Gases. Aeronaut. Eng. Rev., 7, May (1948).

237. MUNSON, T. R., and R. J. SPINDLER: Transient Thermal Behavior of Decomposing Materials. Part I; General Theory and Application to Convective Heating. Presented at the 30th IAS Annual Meeting, New York, January, 1962.

238. MUNTZ, E. P.: Direct Measurement of Velocity Distribution Functions. Rarefied Gas Dynamics, II, New York: Academic Press, 1966.

239. MUNTZ, E. P.: Molecular Velocity Distribution-Function Measurements in a Flowing Gas. Phys. Fluids, II, 64 (1968).

240. MUNTZ, E. P.: Static Temperature Measurements in a Flowing Gas. Phys. Fluids, 5, 1 (1962).

241. MUNTZ, E. P., and S. ABEL: The Direct Measurement of Static Temperatures in Shock Tunnel Flows. Presented at Third Hypervelocity Techniques Symposium, Denver, Colo., March, 1964.

242. MUNTZ, E. P., S. J. ABEL, and F. L. MAGUIRE: The Electron Beam Fluorescence Probe in Experimental Gas Dynamics. Supplement to IEEE Transactions on Aerospace 210, June, 1965,

243. MUNTZ, E. P., and D. J. MARSDEN: Electron Excitation Applied to the Experimental Investigation of Rarefied Gas Flows. Rarefied Gas Dynamics, II, New York: Academic Press, 1963.

244. MUNTZ, E. P., and E. J. SOFTLEY: An Experimental Study of Laminar Near Wakes. TIS R 65 SD 6, General Electric Co., April, 1965.

245. MUNTZ, E. P., and E. J. SOFTLEY: A Study of Laminar Near Wakes. AIAA Journal, 6, 961 (1966).

246. NOLAN, E. J., and S. M. SCALA: The Aerothermodynamic Behavior of Pyrolytic Graphite During Sustained Hypersonic Flight. ARS Conference on Lifting Reentry Vehicles, Palm Springs, California, Paper No. 1969-61, 1961.

247. O'CONNER, T. J., E. G. COMFORT, and L. A. CASS: Turbulent Mixing of an Axisymmetric Jet of Partially Dissociated Nitrogen With Ambient Air. AIAA Journal, 4, November (1966).

248. O'HALLORAN, G. J., and L. WALKER: Determination of Chemical Species Prevalent in a Plasma Jet. ASD-TDR-62-644, U.S. Air Force, 1964.

249. OLSEN, H. N.: Determination of Properties of Optically Thin Argon Plasma. Paper 60, Temperature, Its Measurement and Control in Science and Industry, 2, New York: Reinhold Publishing Company, 1962.

250. OLSON, R. A., and E. C. LARY: Conductivity Probe Measurements in Flames. ARS Paper No. 2592-62, Presented at Ions in Flames and Rocket Exhausts Conference, California, October 10—12, 1962.

251. OSTER, L.: Plasma Diagnostics with the Aid of Cyclotron Radiation. Paper 68, Temperature, Its Measurements and Control in Science and Industry, 2, New York: Reinhold Publishing Company, 1962.

252. PANNABECKER, C.: Inviscid Pressure Drag Coefficients for Sharp and Blunt Cones. RAD-S 210-TR-510-3-4, June, 1965.

253. PAPPAS, C. C.: Effect on Injection of Foreign Gases on the Skin Friction and Heat Transfer of the Turbulent Boundary Layer. LAS Rep. 59—78, January, 1959.

254. PEARCE, W. J.: Plasma-Jet Temperature Measurement. Optical Spectrometric Measurements of High-Temperatures, Univ. of Chicago, 1961.

255. PENNER, S. S.: Quantitative Molecular Spectroscopy and Gas Emissivities. Reading, Mass.: Addison Wesley Publishing Company, 1959.

256. PERRY, H., H. ANDERSON, and F. MIHALOV: Behavior of Reinforced Plastics Surfaces in Contact with Hot Gases. Presented at 15th Annual SPE Tech. Conf., January, 1959.

257. PETRIE, S. L.: An Electron Beam Device for Real Gas Flow Diagnostics. ARL 65-122, U.S. Air Force, 1965.

258. PETRIE, S. L.: Boundary Layer Studies in an Arc-Heated Wind Tunnel. Fn. Report RF Project 2033, Ohio State Univ. Research Foundation, April, 1968.

259. PETRIE, S. L.: Density Measurements with Electron Beam. AIAA Journal, 4, 1680 (1966).

260. PETRIE, S. L.: Electron Beam Diagnostics. AIAA Paper 66-747, September, 1966.

261. PETRIE, S. L.: Flow Field Analyses in a Low Density Arc-Heated Wind Tunnel. 1965 Heat Transfer and Fluid Mechanics Institute Proceedings, Stanford University Press, 1965.

262. PETRIE, S. L., and A. A. BOIARSKI: The Electron Beam Diagnostic Technique for Rarefied Flows at Low Static Temperature. Presented at Rarefied Gas Dynamics Sixth Symposium, 1968.

263. PETRIE, S. L., and G. A. PIERCE: Boundary Layer Studies in Rarefied Plasma Flows. Rarefied Gas Dynamics, II, 1107, New York: Academic Press, 1967.

264. PETRIE, S. L., G. A. PIERCE, and E. S. FISHBOURNE: Analysis of the Thermochemical State of an Expanded Air Plasma. AFFDL-TR-64-191, U.S. Air Force, 1965.

265. PETSCHEK, H., and S. BYRON: Approach to Equilibrium Ionization Behind Strong Shock Waves in Argon. Ann. Phys. NY, 1 (1957).

266. PETTERSON, G., and D. SCHMIDT: A Critical Review of Methods for Determining Properties of Reinforced Plastics at Elevated Temperatures. Presented at 15th Annual SPE Tech. Conf., January, 1959.

267. PFAHL, R. C., JR.: The Determination of Thermal Properties for Charring Ablators from Transient Internal Temperature Measurements. R 720-65-179, AVCO/RAD, January 3, 1966.

268. PLANT, H., and M. GOLDSTEIN: Plastics for High-Temperature Thermal Barriers. Presented at 16th Annual SPE Meeting, Sec. 18-D, February, 1960.

269. POTTER, J. L., G. D. ARNEY, M. KINSLOW, and W. H. CARDEN: Gasdynamic Diagnosis of High-Speed Flows Expanded from Plasma States. IEEE Transactions on Nuclear Science, NS-11, January, 1964.

270. POWERS, D., and J. DICKSON: Investigation of Feasibility of Utilizing Available Heat Resistant Materials for Hypersonic Leading Edge Applications. Bell Aerosystems Co., December, 1960.

271. PREDVODITELEV, A. S.: Studies in Physical Gas Dynamics. NASA-TT-F-505, National Aeronautics and Space Administration, 1968.

272. PROSEN, S., M. KINNA, and F. BARNET: The Development of a Reliable Insulation for Solid Propellant Rocket Motors. Presented at 17th Annual SPE Tech. Conf., **VII**, January, 1961.

273. PYTTE, A., and A. R. WILLIAMS: On Electrical Conduction in a Nonuniform Helium Plasma. Aeronautical Research Lab., U.S. Air Force, 1963.

274. QUINVILLE, J. A., and J. SOLOMON: Ablating Body Heat Transfer. SSD-TDR-63-159, Aerospace Corporation, 15 January, 1964.

275. RAEZER, S. D., and H. L. OLSEN: The Intermittent Thermometer: A New Technique for the Measurement of Extreme Temperature. Temperature — Its Measurement and Control in Science and Industry, **3**, New York: Reinhold, 1962.

276. RAGENT, B., and C. NOBEL: X-ray Densitometer. Vidya Report No. 71, 1962.

277. RAUSCHER, M.: Introduction to Aeronautical Dynamics New York: J. Wiley and Sons, 1953.

278. REBROV, A. K., and R. G. SHARAFUTDINOV: The Structure of the Freely Expanding Carbon Dioxide Jet in Vacuum. Book of Abstracts, **I**, Sixth International Symposium on Rarefied Gas Dynamics, 1968.

279. REED, T.: Induction Coupled Plasma Torch. Massachusetts Inst. of Technology, October, 1960.

280. ROBBEN, F., and L. TALBOT: An Experimental Study of the Rotational Distribution of Nitrogen in a Shock Wave. AS-65-6, University of California, 1965.

281. ROBBEN, F., and L. TALBOT: Measurement of Shock Wave Thickness by the Electron Beam Fluorescence Method. AS-65-4, University of California, 1965.

282. ROBBEN, F., and L. TALBOT: Some Measurements of Rotational Temperatures in a Low Density Wind Tunnel Using Electron Beam Fluorescence. AS-65-5, University of California, 1965.

283. ROMIG, M.: Conical Flow Parameters for Air in Dissociation Equilibrium. Convair Report 7, General Dynamics, May, 1960.

284. ROSENBERRY, J., H. SMITH, and J. WURST: Evaluation of Materials Systems for Use In Extreme Thermal Environments Utilizing an Arc-Plasma-Jet. WADD TR 60-926, U.S. Air Force, 1961.

285. ROSNER, D. E.: Application of Heat Flux Potentials to the Calculation of Convective Heat Transfer in Chemically Reacting Gases. Report No. TP-20, AeroChem Research Laboratories, December 14, 1960.

286. ROSNER, D. E.: Catalytic Probes for the Determination of Atom Concentrations in High Speed Gas Streams. J. Amer. Rocket Soc., **32**, July (1962).

287. ROSNER, D. E.: Diffusion and Chemical Surface Catalysis in a Low Temperature Plasma Jet. ASME Paper No. 61-WA-160. Presented at Winter Annual Meeting, December 19, 1961.

288. ROSNER, D. E.: Diffusion and Chemical Surface Catalysis in a Low-Temperature Plasma Jet. J. Heat Transfer, **84**, No. 4 (1962).

289. ROSNER, D. E.: On the Effects of Diffusion and Chemical Reaction in Convective Heat Transfer. Report No. TM-13, AeroChem Research Laboratories, June 8, 1959.

290. ROSNER, D. E.: Sensitivity of a Downstream Langmuir Probe to Rocket Motor Chamber Conditions. Report No. TP-109, AeroChem Research Laboratories, January, 1965.

291. ROSNER, D. E.: Similitude Treatment of Hypersonic Stagnation Heat Transfer. J. Amer. Rocket Soc., **29**, February (1959).

292. ROSNER, D. E., A. FONTIJN, and S. C. KURZIUS: Chemical Scavenger Probes in Non-equilibrium Gasdynamics. AIAA Journal, **2** (1964).

293. ROTHE, D. E.: Electron Beam Studies of the Diffusive Separation of Helium-Argon Mixtures. Phys. Fluids, **9**, 1643 (1966).

294. ROTHE, D. E.: Flow Visualization Using a Traversing Mechanism. AIAA Journal, **3**, 1945 (1965).

295. RUBESIN, M. W.: A Modified Analogy for the Compressible Turbulent Boundary Layer on a Flat Plate. NACA TN 2917, National Advisory Committee for Aeronautics, 1953.

296. RUBESIN, M. W., C. C. PAPPAS, and A. F. OKUNO: The Effect of Fluid Injection on the Compressible Turbulent Boundary Layer — Preliminary Tests on Transpiration Cooling of a Flat Plate at $M = 2.7$ with Air as the Injected Gas. NACA RM A 55119, National Advisory Committee for Aeronautics, 1955.

297. RUSSELL, G. R., and S. BRYON: Performance and Analysis of a Crossed-Field Accelerator. 63 005, Philco, March, 1963.

298. SANDORFF, P. E.: Orbital and Ballistic Flight; An Introduction to Space Technology. Dept. of Aeronautics and Astronautics, MIT, Cambridge, Mass., 1960.

299. SCALA, S.: The Hypersonic Environment. Aerospace Eng., **22**, (1935).

300. SCHMIDT, D.: Ablative Materials. ASD TR 61-322, U.S. Air Force, 1961.

301. SCHMIDT, D.: Behavior of Plastics in Re-entry Environments. Modern Plastics, **3** (1960).

302. SCHNEIDER, P. J., and R. E. MAURER: Coolant Starvation in a Transpiration — Cooled Hemispherical Shell. J. Spacecraft and Rockets, **5**, June (1968).

303. SCHOPPER, E., and B. SCHUMACHER: Messung von Gasdichten mit Korpuskularstrahlsonden. Z. Naturforsch., **6 a** (1951).

304. SCHURMAN, E. E. H.: Engineering Methods for the Analysis of Aerodynamic Heating. AVCO/RAD TM-63-68, October, 1963.

305. SCHWARTZ, H., and R. FARMER: Thermal Irradiation of Plastic Materials. WADD TR-60-647, U.S. Air Force, 1960.

306. SCHWARTZ, H., and B. LISLE: Effects of High Intensity Thermal Radiation on Plastic Laminates. Presented at 13th Annual SPI Tech. Conf., February, 1958.

307. SCHWEIGER, R.: High-Intensity Arc-I (Instrumentation). AVCO Corp. (No Date.)

308. SCHYMACHER, B.: Dynamic Pressure Stage Elements for the Projection of Intense Mono-energetic Corpuscular Beams Into Gases at High Pressure. Optik, **10**, 116 (1953).

309. SEBACHER, D. I.: An Electron Beam Study of Vibrational and Rotational Relaxing Flows of Nitrogen and Air. Proceedings of the 1966 Heat Transfer and Fluid Mechanics Institute, Stanford University Press, 1966.

310. SEBACHER, D. I.: Diffusive Separation in Shock Waves and Free Jets of Nitrogen Helium Mixtures. AIAA Journal, **6**, 51 (1968).

311. SEBACHER, D. I.: Flow Visualization Using an Electron-Beam Afterglow in N_2 and Air. AIAA Journal, **4**, 1858 (1966).

312. SEBACHER, D. I.: Primary and Afterglow Emission for Low-Temperature Gaseous Nitrogen Excited by Fast Electrons. J. Chem. Phys., **44**, 4131 (1966).

313. SEBACHER, D. I.: Study of Collision Effects Between the Constituients of a Mixture of Helium and Nitrogen Gases When Excited by a 10 Kev Electron Beam. J. Chem. Phys., **42**, 1368 (1965).

314. SEBACHER, D. I., and R. J. DUCKETT: A Spectrographic Analysis of a 1-Foot Hyper-sonic-Arc-Tunnel Airstream Using an Electron Beam Probe. NASA TR-114, National Aeronautics and Space Administration, 1964.

315. SHEAHAN, T. P., and M. CAMAC: Electron Beam Measurements On-Board Re-entry Vehicles. Presented at Rarefied Gas Dynamics Sixth Symposium, 1968.

316. SHERIDAN, R. A., N. S. DIACONIS, and W. R. WARREN: Performance of Several Ablation Materials in Simulated Planetary Atmospheres. R 63 SD 35, General Electric Co., April, 1963.

317. SHERMAN, F. S.: New Experiments on Impact Pressure Interpretation in Supersonic and Subsonic Rarefied Airstreams. Engineering Report No. HE 150-99, Univ. of California, December 21, 1951.

318. SHERMAN, M. P., and J. GREY: Calculation of Transport Properties for Mixtures of Helium and Partly-Ionized Argon. Report No. 673, Princeton University Aeronautical Engineering Laboratory, December, 1963.

319. SHERMAN, M. P., and J. GREY: Interactions Between a Partly-Ionized Laminar Subsonic Jet and a Cool Stagnant Gas. Princeton University, September, 1964.

320. SHIRLEIGH, S.: The Determination of Flame Temperatures by Infrared Radiation. Paper 62, Third Symposium on Combustion and Flame and Explosion Phenomena, 1949.

321. SIMMONS, F. S., and G. E. GLAWE: Theory and Design of a Pneumatic Temperature Probe and Experimental Results Obtained in a High-Temperature Gas Stream. NACA TN 3893, National Advisory Committee for Aeronautics, 1957.

322. SMITH, R. B.: N_2 First Negative Band Broadening Due to Electron Beam Excitation. Presented at Rarefied Gas Dynamics Sixth Symposium, 1968.

323. SMITH, R. B.: Shock Structure in a Diatomic Gas. Presented at Rarefied Gas Dynamics Sixth Symposium, 1968.

324. SNODDY, W., and E. MILLER: Areas of Research on Surfaces for Thermal Control. NASA Washington Symposium on Thermal Radiation of Solids, National Aeronautics and Space Administration, 1965.

325. SOFTLEY, E. J.: Use of a Pulse Heated Fine Wire Probe for the Measurement of Total Temperature in Shock Driven Facilities. AIAA Paper No. 68-393, April, 1968.

326. SOFTLEY, E. J., E. P. MUNTZ, and R. E. ZEMPEL: Experimental Determination of Pressure, Temperature, and Density in Some Laminar Hypersonic Near Wakes. TIS-R 64 SD 35, General Electric Co., 1964.

327. SPRENGEL, U.: Kalorimetrische Messungen von örtlichen Temperaturen in einem Stickstoff-Plasmastrahl. Raumfahrtforschung und Technik, Beilage zur Atompraxis, 1966.

328. STAATS, G. E., W. K. McGREGOR, and J. P. FROLICH: Magnetogasdynamic Experiments Conducted in a Supersonic Plasma Arc Tunnel. AEDC TR-67-266, U.S. Government, 1968.

329. STALDER, J. R., F. K. GOODWIN, B. RAGENT, and C. E. NOBLE: Aerodynamic Application of Plasma Wind Tunnels. WADD TN 60-1, U.S. Air Force, 1960.

330. STEG, L.: Materials for Heat Protection of Satellites. J. Amer. Rocket Soc., 9 (1960).

331. STEIN, E.: Summary of AVCO RAD Studies on the Use of Effective Properties for the Thermal Analysis of Heat Shield Materials. AVCO/RAD, Wilmington, Mass., March 16, 1964.

332. STERN, M. O., and E. M. DACUS: Piezoelectric Probe for Plasma Research. Rev. Sci. Instr., 32, No. 2, February (1961).

333. STEVEN, G.: High-Temperature Thermocouples. High-Temperature Technology, New York: John Wiley and Sons, 1956.

334. STEWART, J. D.: Transpiration Cooling: An Engineering Approach. MSVD-TIS-R 59 SD 338, General Electric Co., May 1, 1959.

335. STROUBAL, G., D. M. CURRY, and J. M. JANNEY: Thermal Protection System Performance of the Apollo Command Module. Presented at AIAA/ASME 7th Structures and Materials Conference, Cocoa Beach Florida, April 18—20, 1966.

336. SU, C. H., and S. H. LAM: Continuous Theory of Spherical Electrostatic Probes. Phys. Fluids, 6, October (1963).

337. Suits, G. C.: The Determination of Arc Temperatures from Sound Velocity Measurement. Phys., **6** (1935).

338. Sutton, G.: Ablation of Reinforced Plastics in Supersonic Flow. J. Aerospace Sci., **6** (1960).

339. Talbot, L.: Theory of the Stagnation-Point Langmuir Probe. Phys. Fluids, **3**, 289—297 (1960).

340. Talbot, L.: Theory of the Stagnation-Point Langmuir Probe. Tech. Report HE-150-168, University of California, March 30, 1959.

341. Talbot, L., J. E. Katz, and C. L. Brundin: A Comparison Between Langmuir Probe and Microwave Electron Density Measurements in an Arc-Heated Low-Density Supersonic Wind Tunnel. Engineering Report HE-150-186, University of California, January 27, 1961.

342. Tanaki, F., and C. S. Sim: Flash X-Ray Radiography for the Density Measurement in a Hypersonic Air Flow. J. Phys. Soc. of Japan, **14**, No. 5 (1959).

343. Tempelmeyer, K. E.: Development of a Steady-Flow J X B Accelerator for Wind Tunnel Application. TDR-64-261, U.S. Air Force, 1964.

344. Tillian, D. J.: A Survey of Plasma Arc Heaters. Report 00-49, Vought Astronautics Division, Ling-Temco-Vought Company, April 18, 1962.

345. Tirumalesa, D.: A Preliminary Study of the Flow Field Around a Blunt Body in Hypersonic Low Density Flow. USXAW 105, Univ. of Southern Calif., November, 1965.

346. Tirumalesa, D.: An Experimental Study of Hypersonic Rarefied Flow Over a Blunt Body. AIAA Journal, **6**, 369 (1968).

347. Tolosko, R. J.: A Study of the Flow Field, Aerodynamic Heating, and Ablation of Sharp Cones Entering the Earth's Atmosphere at Supersonic Velocity. AVCO/RAD TM-62-17, April 18, 1962.

348. Trochan, A. M.: Measurements of Parameter for Gas Flows by Means of a Beam of East Electrons. J. Appl. Math. and Phys., **3**, 81 (1964).

349. Vassallo, F. A.: A Fast Acting Miniature Enthalpy Probe. AIAA Paper No. 68-391, April, 1968.

350. Vassallo, F. A.: Miniature Enthalpy Probes for High-Temperature Gas Streams. ARL 66-0115, U.S. Air Force, 1966.

351. Vassallo, F. A., and J. Beal: Structural and Insulative Characteristics of Reinforced Plastic Materials During Ablation. Presented at 16th Annual SPI Tech. Conf., February, 1961.

352. Vassallo, F. A., N. Wahl, G. Sterbutzel, and J. Beal: The Study of Ablation of Structural Plastic Materials. WADC TR 59-368, Part 1, U.S. Air Force, 1959.

353. Vidale, G. L.: Measurement of the Absorption of Resonance Lines. GE-R 60 SD 331, General Electric Space Sciences Lab., May, 1960.

354. Vidale, G. L.: Measurement of the Vapor Pressure of Atomic Species from Spectrophotometric Measurements of the Absorption of the Resonance Lines. IV GE-R 60 SD 333, General Electric Space Sciences Lab., July, 1960.

355. Vidale, G. L.: Measurement of the Vapor Pressure of Atomic Species from Spectrophotometric Measurements of the Absorption of the Resonance Lines. III., GE-R 60 SD 390, General Electric Space Sciences Lab., June, 1960.

356. Vogan, J.: Thermal Protective Surfaces for Structural Plastics. WADD TR 60-110, Part 1, U.S. Air Force, 1960.

357. Wada, I.: Experimental Study of Hypersonic Low Density Flow by Using the Electron Beam Fluorescence Method. Presented at Rarefied Gas Dynamics Symposium, 1967.

358. Wainwright, J. B.: Experimental Investigation of the Density Flow Field in the Stagnation Region of a Blunt Body in High Speed Rarefied Flow. BD 339-101, CELESCO, 1966.

359. WALLACE, J. E.: Hypersonic Turbulent Boundary Layer Measurements Using an Electron Beam. AN-2112-Y-1, Cornell Aeronautical Laboratory, August, 1968.

360. WARREN, W., and N. DIACONIS: The Performance of Ablation Materials as Heat Protection for Re-Entering Satellite. WADD TR 60-58, U.S. Air Force, 1960.

361. WAYMOUTH, J. F.: Perturbation of a Plasma by a Probe. Phys. Fluids, 7, 1843—1854 (1964).

362. WEINSTEIN, L. M., R. D. WAGNER JR., and S. L. OCHELTREE: Electron Beam Flow Visualization in Hypersonic Helium Flow. AIAA Journal, 6, 1623 (1968).

363. WELCH, N. E.: The Laser Anemometer — A New Tool In Flow Measurement. Space Age News, 12, March (1969).

364. WELCH, N. E., and W. J. TOMME: The Analysis of Turbulence from Data Obtained with a Laser Velocimeter. AIAA Paper No. 67-179, January, 1967.

365. WETHERN, R. J.: Method of Analyzing Laminar Air Arc-Tunnel Heat Transfer Data. AIAA Journal, I, July (1963).

366. WHARTON, C. E.: Plasma Diagnostics. AIAA Paper No. 63-367, Fifth Biennial Gas Dynamics Symposium, Northwestern University, August 14—16, 1963.

367. WIEBELT, PARMA, and KNEISSL: Spacecraft Temperature Control by Thermostatic Fins-Analysis, Part II. NASA CR-155, National Aeronautics and Space Administration, January, 1965.

368. WILDHACK, W. A.: A Versatile Pneumatic Instrument Based on Critical Flow. Review of Sci. Instr., 21, January (1950).

369. WILLIAMS, P. M., and J. GREY: Simulation of Gaseous Core Nuclear Rocket Characteristics Using Cold and Arc Heated Flows. NASA Contractor Report No. CR-690, National Aeronautics and Space Administration, June, 1967.

370. WINKLER, E. L., and R. GRIFFIN JR.: Effects of Surface Recombination on Heat Transfer to Bodies in a High Enthalpy Stream of Partially Dissociated Nitrogen. NASA TND-1146, National Aeronautics and Space Administration, 1961.

371. WINKLER, W.: Interferometry in Physical Measurements in Gas Dynamics and Combustion. Princeton, N. J.: Princeton University Press, 1954.

372. WIZANSKY, D., and E. RUSS: An Oxyacetylene Flame Apparatus for Surface Ablation Studies. Tech. Report HE-150-167, Univ. of California, 28 January, 1959.

373. WIZANSKY, D., E. RUSS, and W. GIEDT: An Oxyacetylene Flame Apparatus for Surface Ablation Studies. Tech. Report HE-150-171, Univ. of California, May, 1959.

374. WURST, J., J. CHERRY, D. GERDEMAN, and N. HECHT: The High-Temperature Evaluation of Aerospace Materials. AFML-TR-66-308, U.S. Air Force, 1966.

375. WURST, J., J. CHERRY, and N. HECHT: The Evaluation of High Temperature Materials. QPR No. 1, University of Dayton Research Institute, April, 1964.

376. WYATT, L. A.: Low-Speed Measurements of Skin Friction on a Slender Wing. ARC-R & M-3499, Aeronautical Research Council, 1968.

377. YARGER, F.: Electromechanical Influences on High-Temperature Flows. ATL TR-66-38, U.S. Air Force, 1966.

378. YASUBARA, M.: Rotational Temperature Measurements of the Flow Expanding From a Low Density Sonic Orifice. USCAE 103, Univ. of Southern California, August, 1966.

379. YEH, Y., and H. Z. CUMMINS: Localized Fluid Flow Measurements with an He-Ne Laser Spectrometer. Applied Physics Letters, 4, May, 1964.

380. ZEMPEL, R. E., and E. P. MUNTZ: Slender Body Near Wake Density Measurements at Mach Numbers Thirteen and Eighteen. TIS-R 63 SD 55, General Electric Co., 1963.

381. ZLOTNICK, M., and B. NORDQUIST: Calculation of Transient Ablation. Presented at International Heat Transfer Conference, Boulder, Colorado, September, 1961.

Subject Index

Printer: R. Spies & Co., A-1050 Wien